ROUTLEDGE LIBRARY EDITIONS:
LOGIC

Volume 22

THE TRADITIONAL
FORMAL LOGIC

T0298714

THE TRADITIONAL FORMAL LOGIC

A Short Account for Students

WILLIAM ANGUS SINCLAIR

Routledge
Taylor & Francis Group

LONDON AND NEW YORK

First published in 1937 by Methuen & Co Ltd
Fifth edition published 1951

This edition first published in 2020
by Routledge
2 Park Square, Milton Park, Abingdon, Oxon OX14 4RN

and by Routledge
52 Vanderbilt Avenue, New York, NY 10017

Routledge is an imprint of the Taylor & Francis Group, an informa business

© 1937, 1951 William Angus Sinclair

British Library Cataloguing in Publication Data
A catalogue record for this book is available from the British Library

ISBN: 978-0-367-41707-9 (Set)
ISBN: 978-0-367-81582-0 (Set) (ebk)
ISBN: 978-0-367-42262-2 (Volume 22) (hbk)
ISBN: 978-0-367-42629-3 (Volume 22) (pbk)
ISBN: 978-0-367-85434-8 (Volume 22) (ebk)

Publisher's Note
The publisher has gone to great lengths to ensure the quality of this reprint but points out that some imperfections in the original copies may be apparent.

Disclaimer
The publisher has made every effort to trace copyright holders and would welcome correspondence from those they have been unable to trace.

The Traditional
Formal Logic

A SHORT ACCOUNT FOR STUDENTS

by

WILLIAM ANGUS SINCLAIR

Late Reader in Philosophy
in the University of Edinburgh

METHUEN & CO LTD

11 NEW FETTER LANE · LONDON · EC4

TO

MY MOTHER

First Published November 11th 1937
Second Edition, Revised, March 1945
Third Edition March 1947
Fourth Edition, Revised, July 1949
Fifth Edition January 1951
Reprinted five times
Reprinted 1965

5.7

CATALOGUE NO. 02/3884/41

PRINTED AND BOUND IN GREAT BRITAIN BY
BUTLER AND TANNER LTD, FROME AND LONDON

PREFACE TO THE FIFTH EDITION

THIS book was written because, like many teachers of philosophy, I had long felt with increasing dissatisfaction that it was a waste of time and opportunity to spend lecturing or tutoring hours on a subject as simple and straightforward as the outlines of the traditional formal logic. This is therefore intended as an elementary book which the ordinarily competent student can be left to work through by himself. It is meant to lead the unassisted reader by stages to the degree of familiarity with the subject that is described in the Introduction as a necessary preparation for further study, so that the teacher who asks his pupils to read it can reasonably expect them to acquire thereby sufficient knowledge of formal logic to understand the discussion of advanced logical topics. I hope that the pupils will not feel any serious breach of continuity whatever be the teacher's own position and manner of treatment ; whether, for instance, he adds the further details of formal logic if he thinks them important, or develops symbolic logic, or criticizes the epistemological presuppositions on which a formal logic is said to rest, or treats the whole as an historically important misconception, or deals with the subject in some other way of his own. It has therefore become our practice at Edinburgh to ask students to work through this book in the long vacation prior to entering the Class of Philosophy.

The book is intended also for those who want to read logic for the gain in understanding mentioned in the Introduction, or to satisfy some professional requirement such as Teaching Certificate or law examinations.

I am indebted principally to Professor Norman Kemp Smith for innumerable suggestions and improvements, and in many other ways also. In addition Professor A. J. D. Porteous, Professor James Drever and my mother were kind enough to read the whole, and Father Ian Ross to read parts, of the manu-

v

script and to make many comments and corrections. Professor Porteous and my mother assisted in reading the proofs.

In the Second and Third Editions there were no substantial changes, but the opportunity was taken to incorporate suggestions made by Professor T. E. Jessop, Mr. George Brown, Mr. D. R. Cousin, Professor W. H. F. Barnes and others. In the Fourth and Fifth Editions, suggestions by the Rev. Alan Fairweather, the Rev. Geddes MacGregor, and particularly by Miss M. J. Levett, have been adopted. The principal of these are amplifications in the treatment of rules of syllogism (page 51) and of disjunctives (page 74 *et seq.*).

W. A. S.

July 1950

CONTENTS

INTRODUCTION

WHAT logic is, and why it is worth studying, are difficult to understand until we are reasonably familiar with the variety of it known as the traditional formal logic.

Just as there are many different 'philosophies', and many different theories in any controversial field, so there are many different 'logics'. The traditional formal logic is only one among these numerous alternatives. It has, however, a unique importance, for it is a straightforward and comprehensive body of doctrine that has been taught as an essential of the higher education in Western Europe from the twelfth century to the present day. It was originated by Aristotle more than two thousand years ago, was partially neglected in the period after the decay of ancient learning, and came to its full influence on the rediscovery of the bulk of his writings about 1150. With grammar and rhetoric it formed the *trivium*, the first three of the seven liberal arts of the medieval universities, and it is still taught substantially unchanged in the universities of our own time.

The technical terms and notions which it employs passed long ago into the languages and thought of the Western world, and are so influential there that ignorance of it hinders the understanding of much in our civilization that would otherwise be plain. This in itself makes some knowledge of the traditional logic an educational essential.

Moreover, all the other logical theories that have been and are being advanced have arisen out of the traditional doctrine, either by extension of it, or by disagreement with it varying from minor criticism to total rejection. The student must acquaint himself with that doctrine before he can profitably discuss the nature of logic and the other questions that will in the usual order of philosophical study be brought to his notice, such as questions of the nature of knowledge, of 'modern logic', and of the developments connecting the logic of medieval times with the scientific method of to-day.

The surpassing importance of the traditional formal logic is well shown by the common use of the name ' logic ' by itself to mean the traditional doctrine, much as the name ' geometry by itself is commonly used to mean the traditional or Euclidean geometry, which likewise is only one of many alternatives.

This short book is intended to give an account of that traditional doctrine, or more exactly an account of those fundamental parts of it that a man must know if he is to appreciate what has for so long been an important element in our common intellectual inheritance and if he is to understand the later developments in logic and philosophy.

The ingenuity of medieval and of some later logicians has from time to time embellished the tradition with further complications, and although only a few of these have survived, the present state of the tradition is complex and detailed far beyond what is profitable for any but the specialist to study.   Only the essentials are dealt with here, but they form the necessary minimum with which the student must be not only acquainted but familiar.   In blunter language, students who do not read and understand the whole of the book, footnotes and all, and who attempt to make selections for themselves, are likely to be unable to answer the sort of questions usually set in elementary logic examinations.   This kind of logic must therefore be learned and assimilated much as we learn and assimilate, not very critically, the grammar of our own or of a foreign tongue.   It is, incidentally, much easier than students sometimes anticipate.

This book has been designed for consecutive reading, and the student should refrain from passing to a new point until he clearly understands the point with which he is dealing.   Cursory reading, or reading in another order, is likely to be unprofitable, as the later stages are intelligible only in the light of the earlier.

Bearing these considerations and purposes in mind, we can now set about familiarizing ourselves with the traditional formal logic.   Doubts on the doctrines and practices involved will occur to most readers, but these had better be reserved for the time being, as the purpose of this book is the strictly limited one of making the tradition familiar, and not of appraising or developing it.   For the latter types of discussion the reader can turn later to books dealing specifically with them.

CHAPTER I

PROPOSITIONS

LOGIC, we all know, is a science or inquiry dealing with topics such as reasoning, inferring, and arguing, and especially with reasoning, inferring, and arguing aright. We may find it difficult to give an adequate definition of 'logic' and 'logical', but we feel confident that these words are properly used in sentences such as *He is a clear and logical thinker*, or *That may be a logical conclusion but I don't agree with it*, or *There is no regard for logic in your argument*. All we can at present say is that logic is in some sense an examination of argument, statement, implication, inference, and the like.[1] It is important to understand that the traditional logic confines itself to statements and arguments that can be true or false, valid or invalid, and that it does not deal with the many other forms of speech that make no claim to be either true or false, such as commands, wishes, ejaculations, and so forth. In other words, it confines itself to sentences that have verbs in the indicative mood, or that can be restated as sentences having verbs in the indicative mood.

[1] It is important to understand that we may be able to argue and infer and conclude quite correctly without knowing anything about logic. There is often confusion on this point, and it is sometimes even alleged that a man ignorant of logic is on that account unlikely to be accurate in his thinking.

This confusion disappears on making clear the distinction between the carrying out of an activity and the study of that activity. The activity of a living organism, namely being alive, is different from the study of it, which is physiology ; similarly the activity of talking or writing grammatically is different from the study of it, which is grammar ; and similarly the activity of thinking and arguing logically is different from the study of it, which is logic.

Old Parr of Banbury lived for more than a hundred and fifty years though he knew no physiology ; Homer wrote grammatically though he had never heard of grammar ; and we can argue logically without knowing any logic.

Whether the study of an activity affects the efficiency with which we perform it, and in what way and how far, is another question.

3

Consider examples :

> *Some flowers are scarlet*
> *All men are moralists*
> *A wanderer is man from his birth*
> *Not all stocks and shares are worth having*
> *If supplies increase, prices fall*
> *That is either a horse or a mule*
> *It must have been raining for the pavements are wet.*

The first point to be noticed is that these differ considerably in their degree of complexity, for the latter examples are comparatively involved and complicated in structure, whereas the first two are comparatively simple. It will be seen that the simplest kinds of statement to be found are similar in their structure to these first two examples,

> *Some flowers are scarlet*
> *All men are moralists.*

Statements of this simple kind are known as **propositions,** and as such are of fundamental importance for our present purpose, because the traditional logic maintains the convention that all statements and arguments, no matter how involved and complicated, can be analysed and shown to consist either of simple propositions, or of groups of simple propositions standing in some systematic relation to each other. Though we have not yet studied the structure of propositions, we can with moderate facility recognize to be propositions such examples as the two previously quoted, and others similarly simple.

We shall now investigate the nature and structure of propositions, but before doing so we must be clear on a most important point, namely that the truth or falsity of the propositions is at this stage irrelevant, for what is being examined is the *structure* of the propositions and not the truth or falsity of what they assert. This may be made clearer by reference to the similar conditions in the study of grammar, for the grammatical and syntactical structure of a sentence is independent of its truth or falsity. For instance, the word *efficiently* in the sentence *This business is efficiently run* is an adverb, and it remains an adverb whether the business in question is in fact capably managed or incompetently mismanaged. In a similar way, the truth or falsity of a proposition is irrelevant to a study of its structure. In the technical language of logic, it is the **form** of the proposition that is important for our purpose and not its matter.

STRUCTURE OF THE PROPOSITION

All propositions may be regarded as exhibiting a definite structure. Consider examples, such as

> *Some trees are conifers*
> *Whales are not fish.*

In these propositions there is, first, something spoken about, the **subject** of the discussion, namely *trees* or *whales*. Secondly, there is something said about the subject, 'predicated' of it, and therefore called the **predicate.**

The traditional doctrine insists on the convention of treating all propositions as stating a relation between two classes of entities, between a **subject** and a **predicate.**[1] Thus the proposition *Some trees are conifers* is treated as asserting that a certain relation holds between the class of entities called *trees*, which is the subject of the proposition, and the class of entities called *conifers*, which is the predicate of the proposition. In the second example a relation of a somewhat different kind is asserted to hold between *whales* which is the subject, and *fish* which is the predicate.

So there are three factors in every proposition. First, there is a class of entities spoken about, namely the subject ; secondly, there is another class of entities, namely the predicate ; and thirdly, there is the relation stated to hold between these two classes. This relation is indicated by a part of the verb *to be*

---

[1] A more detailed analysis shows that this involves *two* distinguishable conventions :

(a) The convention of treating a proposition as stating a relation between its Subject and its Predicate.

(b) The convention of treating Subject and Predicate as classes of entities.

(e.g. *are* or *are not*) in conjunction with an adjective of quantity (e.g. *all* or *some*) which is prefixed to the subject.

The subject and predicate are called technically the **terms** of the proposition, and the part of the verb *to be* is called the **copula.** A proposition thus consists of two terms, namely the subject term and the predicate term, (commonly called more briefly the subject and the predicate), and of the copula and the quantitative adjective which serve in conjunction to indicate the relation in which these two terms stand to each other.

Very few of the statements of ordinary speech take forms in which the subject, the predicate and their relation are readily distinguishable, for ordinary speech is in a high degree complex, condensed and elliptical, since most statements in normal language consist not of one proposition, but of several involved one with another in comparatively brief wording. So for the purposes of logical study, (since the traditional logic maintains that any statement however complex consists of these simple units called propositions) the statements of ordinary speech must be analysed, and the constituent propositions disentangled from one another and stated separately. They must be stated separately and in proper **logical form** as it is called, that is, in such a way that the subject, the predicate, and the relation between them are clearly and unmistakably expressed and distinguished.

The process of restating ordinary speech in logical form must be understood, and practised till it can be done with facility. This may seem dull and sometimes irritating, but it is essential to the appreciation of logical theory, and is useful as a mental discipline. It is a common criticism of a rhetorical controversialist that he 'does not know what he is saying', and the exercise of restating assertions in logical form is important as making clear exactly what has been said in the looser forms of ordinary speech. To turn one's own dicta into logical form is in consequence often illuminating and sometimes humbling. Moreover, by the exercise of restating in logical form, the student is brought to understand the structure of propositions more adequately than by any description of that structure given by another.

The restating of given sentences in logical form in accordance with the traditional logic is best undertaken in three distinct stages. First of all we have to discover what is the subject, that is to say the class of entities that is being talked about. Secondly, we have to discover what is the predicate, that is to say the other class of entities involved. Lastly, after the terms of the proposition have in this way been clearly stated, the relation in which these two terms stand to each other can be noted, and the appropriate copula and quantitative adjective added to indicate this.

In most cases the two terms are fairly readily discoverable, and we shall now consider some typical sentences, confining ourselves for the moment to the first two stages, namely, finding first the subject and then the predicate, leaving the relation between them to be examined later.

In the sentence *Politicians are not always rich men*, it is clear that the subject (the class of entities being talked about) is *politicians*, and that the predicate (the class of entities that is being related to the subject) is *rich men*. Neglecting for the moment the relation in which these classes stand to each other, we have solved the simple problem of finding the terms of the proposition that would be the logical form of the sentence in question.

To take a rather less obvious example, the subject of the sentence *Dogs are faithful* is clearly enough *dogs*, but at first it appears difficult to find the predicate, for we are given not a class of entities, but only a quality *faithful*. The traditional logic treats the sentence *Dogs are faithful* as a colloquial way of saying *Dogs are faithful animals*, which gives us the class *faithful animals* as the predicate. Similarly the predicate in *Men are mortal* is *mortal beings*, and in *Diamonds are very hard* it is *very hard stones* or *very hard things*, or some similar phrase meaning not the qualities in question but entities possessing the qualities.

In more complicated cases the logical subject may not be the grammatical subject of the sentence, and the predicate may not correspond to the grammatical predicate. Very commonly a sentence has to be restated so that the significance of its verb is included in one of the terms. For instance, the logical subject of the sentence *Diesel engines burn heavy oil* is simply *Diesel engines*, but the logical predicate has to include the significance of the verb and therefore it is *engines that burn heavy oil*. Drastic restatement may be necessary, as in *Virtue is not given to all*, where the subject (the class of entities spoken about) is not *virtue* but *persons*, a large class some of whose members are excluded from the other class of *persons gifted with virtue*, which is the predicate.

An even more extreme instance is *Sometimes there is no quorum*, where the subject, or class of entities spoken about (using ' entities ' in the widest possible sense), is *occasions of meeting*, and the predicate is *occasions on which there is no quorum*, so that the logical form of the sentence would be the somewhat strange but quite intelligible one, *Some occasions of meeting are occasions on which there is no quorum*.

Commonly in ordinary discourse only a few words are articulated, for it is assumed that the other words necessary to complete the meaning are known to the listener or reader, and these words must be inserted if the statement is to be expressed in logical form. The single word *Uncut*, uttered on examining new books, is elliptic for the lengthier statement *These books are books whose leaves have not been cut*, of which the subject is *these books* and the predicate is *books whose leaves have not been cut*.[1]

We have now reached the stage of seeing that a proposition consists of two terms, and have learned how to find and express those terms even if they are partially omitted or obscured by the colloquial forms of our normal speech. The next point is naturally the relation between those two terms, (the relation which is indicated by the copula in conjunction with the quantitative adjective prefixed to the subject). We shall now consider what different kinds of relation are recognized by the traditional doctrine.

---

[1] Isolated words that represent propositions must be carefully distinguished from isolated words that are only ejaculations, interjections, commands and the like, for these latter do not represent propositions and are in consequence not dealt with here, for logic is not concerned with them. See page 3.

On inspection it appears in the first place that in any proposition we speak either about the whole of the subject or about a part of it. Any proposition, whatever else it may be, is an assertion either about the whole of the subject or about a part of it at least. This is indicated by words such as *All* and *No* (*None*) as in *All men are liars* or *No ungulates are carnivores*, which are statements about the whole of the subject ; or by words like *Some*, as in *Some men are fortune's favourites* and *Some men are not fortune's favourites*, which are statements about a part of the subject. These two kinds of proposition are distinguished as **universal** propositions and **particular** propositions, a universal proposition being a statement about the whole of the subject, and a particular proposition being a statement about a part of the subject at least.[1] That is to say, we can indicate in a proposition that the predicate class is related either to *all* the entities that compose the subject class, or to *some* at least of them.

The relation between the two terms of a proposition, thus indicated by the prefixed adjective, must then be either universal or particular in **quantity,** as the technical phrase has it, according as the proposition is about the whole of the subject or about a part of it. The relation in which the two terms of a proposition stand to each other must, in quantity, be either universal or particular. In other words, every proposition must be either a universal proposition or a particular proposition.

Further, it appears that in addition to the alternatives in quantity, there are alternatives of another kind also in the relationship of the terms, for in every proposition we assert either that the subject *is* something or other, or that the subject *is not* something or other.[2] That is, we assert either that the subject class *is* included, partially or wholly, in the predicate class ; or we assert that the subject class *is not* included, partially or wholly, in the predicate class. Either we assert that the subject stands in a relation (universal or particular) to the predicate, or we assert that the subject does not stand in that relation

---

[1] See page 12 on *Some*.

[2] This is, of course, not the case in commands, wishes and the like, but it is to be remembered that they are not propositions. See page 3.

2

to the predicate. In other words, the relation between the terms—whether it be universal or particular—is either affirmed or negated. In **quality,** as it is called, every proposition must be either **affirmative** or **negative.** This is indicated by the affirmative or negative form of the copula, e.g. *are* or *are not.*[1]

This means that the terms of a proposition may stand related to each other in one or other of two ways in quantity, and in one or other of two ways in quality. These two pairs of alternatives combine to give four alternatives. These exhaust the possibilities of relationship between subject and predicate. That is, the relationship must be either :

universal and **affirmative**	*All cows are ruminants*
or **universal** and **negative**	*No salts are elements*
or **particular** and **affirmative**	*Some peasants are poets*
or **particular** and **negative**	*Some men are not heroes.*

These four possible relationships of the subject and predicate are indicated by the copula and the prefixed adjective of quantity. The proposition must in consequence take one or other of four possible forms,

All —— *are* —— (The universal affirmative proposition)
No —— *are* —— (The universal negative proposition)
Some — *are* —— (The particular affirmative proposition)
Some — *are not* —— (The particular negative proposition).

Thus the analysis of propositions in the traditional logic, owing to the conventions upon which it rests, results in an extreme simplification by treating all propositions as consisting of two terms related to each other in a way that must be one or other of only four alternatives.[2]

---

[1] Note that, strictly, the two formal factors (universal and particular, affirmative and negative) refer to the proposition as a whole. I.e. it is the *proposition* that is universal or particular, and not the terms ; and it is the *proposition* that is affirmative or negative, and not the relation between the terms.

[2] Reference to the schematic summary on page 97 may be helpful from this point onward.

To express the statements of ordinary speech in logical form is, it now appears, a straightforward procedure, which we can carry out in three separate and comparatively simple steps.

I. We have to discover the **subject,** the class of entities that is being spoken about, and state it in a form of words that is, if need be, more adequate and explicit than the original sentence.

II. We have similarly to discover and state the **predicate,** the class of entities that is said to stand in some specific relation to the subject.

III. We have to discover what that relation is, knowing it must be one or other of the four alternatives, and to indicate it by the appropriate **quantitative adjective** and **copula.** These must in consequence take one or other of the four forms:

$$All \quad \text{——} \quad are \text{——}$$
$$No \quad \text{——} \quad are \text{——}$$
$$Some \text{——} \quad are \text{——}$$
$$Some \text{——} \quad are \ not \text{——}$$

We can now close this chapter, which covers the first major stage in this study of logic, by taking some typical sentences and expressing them in logical form.

EXAMPLES

Some boys aren't interested in games.

I. The Subject is *boys.*   II. The Predicate is *persons interested in games.*

III. The relation between the Subject and the Predicate is such that the proposition is Particular, and Negative.

So we have a particular negative proposition with *boys* as subject, and *persons interested in games* as predicate, thus :

Some boys are not persons interested in games.

Some is interpreted as covering even a single instance, i.e. *some* means *one at least*. It is also to be noted that *some* does not exclude *all*, i.e. *some* means *one at least, possibly more, possibly all, subject to other conditions not here specified*. The statement that some private Banks are Limited Liability Companies is not inconsistent with a subsequent statement that they are all Limited Liability Companies. It is important to understand that in the traditional logic *some* has this conventional usage, whereby it does not exclude *all*.

Golf clubs are sometimes made entirely of metal.

I. The Subject is *golf clubs*.

II. The Predicate is *things made entirely of metal*.

III. The relation between the Subject and the Predicate is such that the proposition is Particular, as is shown by the modifying word *sometimes* ; and it is Affirmative.

So we have a particular affirmative proposition with *golf clubs* as subject, and *things made entirely of metal* as predicate, thus :

Some golf clubs are things made entirely of metal.

The pure bred Cairn never has a timid disposition.

I. The Subject is clearly enough the class *pure bred Cairns* (or *Cairn terriers*), even though the singular number has been used in the sentence.

II. The Predicate is *dogs with timid dispositions.*

III. The relation between the Subject and the Predicate is such that the proposition is Universal, for the assertion is made about all *pure bred Cairns*, about the whole of the Subject class *pure bred Cairns* and not merely about part of the class. It is also Negative.

So we have a universal negative proposition with *pure bred Cairns* as subject, and *dogs with timid dispositions* as predicate, thus :

No pure bred Cairns are dogs with timid dispositions.

A wood pigeon nests high.

I. The Subject is *wood pigeons*, for the singular number of the noun is here again intended to refer to the class and not to any specified member of it.

II. The Predicate is *birds that nest high.*

III. The relation between the Subject and the Predicate is such that the proposition is Universal, as in the previous example, and also Affirmative.

So we have a universal affirmative proposition with *wood pigeons* as subject, and *birds that nest high* as predicate, thus :

All wood pigeons are birds that nest high.

" Ticket-holders only "

This is a very condensed statement. Such a notice is intended to inform readers both that people without tickets are not admitted and also that ticket-holders are or may be admitted. In restating in logical form, these two statements have to be expressed separately, as separate propositions. Take the first of them first, namely that people without tickets are not admitted.

I. The Subject is *persons other than ticket-holders*.
II. The Predicate is *admissible persons*.

III. The relation between the Subject and Predicate is such that the proposition is Universal, and Negative.

So we have a universal negative proposition with *persons other than ticket-holders* as subject, and *admissible persons* as predicate, thus:

No persons other than ticket-holders are admissible persons.

Now take the second statement, namely that ticket-holders are or may be admitted.

I. The Subject is *ticket-holders*.
II. The Predicate is *admissible persons*.

III. The relation between the Subject and the Predicate is such that the proposition is Universal, and Affirmative.

So we have a universal affirmative proposition with *ticket-holders* as subject, and *admissible persons* as predicate, thus:—

All ticket-holders are admissible persons.

Not all verse is poetry.

I. The Subject is *verse compositions*.
II. The Predicate is *poems*.

III. The relation between the Subject and
the Predicate is clearly such that the
proposition is Negative ; but it is *not*
Universal, as one might at a casual glance
suppose, but only Particular, for the sentence does not assert that no verse at all
is poetry, but only that some verse is
not poetry.

So we have a particular negative proposition with *verse compositions* as subject, and *poems* as predicate, thus :

Some verse compositions are not poems.

The words *not all* require care, for they are equivalent to *some are not,* and do not mean *none are.*

All that glisters is not gold.

I. The Subject is *glister-*   II. The Predicate is *golden*
   ing things.   *things.*

III. The relation between the Subject and the
Predicate is the same as in the previous
example. It is clearly such that the
proposition is Negative, and not Universal but Particular, for the sentence
does not mean that no glistering things
at all are gold, but only that some of
them are not gold.

So we have a particular negative proposition with *glistering things* as subject, and *golden things* as predicate, thus :

Some glistering things are not golden things.

Sentences in the form *All —— are not ——* are ambiguous and require care in interpretation. For instance, the statement *All the medical students are not interested in politics* might be intended to mean that none are, or it might be intended to mean only that some of them are not.

Flying fish don't really fly.

I. The Subject is *flying fish*.

II. The Predicate is *beings that really fly* or *beings capable of genuine flight* or a similar phrase.

III. The relation between the Subject and the Predicate is such that the proposition is Negative; and also Universal, as the assertion is made about all flying fish and not merely about some of them.

So we have a universal negative proposition with *flying fish* as subject and *beings capable of genuine flight* as predicate, thus

No flying fish are beings capable of genuine flight.

Now come some more difficult examples to show how the logical form of the proposition may be very unlike the grammatical form of the sentence. The logical subject of the proposition need not be, and indeed seldom is, the same as the grammatical subject of the sentence.

Blessed are the merciful.

I. One might be tempted to say that the Subject is *blessed persons*, but a little consideration shows that it is not *blessed persons* about whom the assertion is being made. The sentence is really an assertion about *merciful persons*, which is the Subject of the proposition.

II. The Predicate is *blessed persons*.

III. The relation between the Subject and the Predicate is clearly such that the proposition is both Universal and Affirmative.

So we have a universal affirmative proposition with *merciful persons* as subject, and *blessed persons* as predicate, thus :

All merciful persons are blessed persons.

There is not a man of them but has his price.

I. The Subject is the men in question, i.e. *men forming that group*, or some such phrase.

II. The Predicate is *men who have their price*.

III. The plurality of negative words makes the nature of the relation between the Subject and the Predicate a little obscure at first, but it is seen to be such that the proposition is both Universal and Affirmative.

So we have a universal affirmative proposition with *men forming that group* as subject, and *men who have their price* as predicate, thus :

All men forming that group are men who have their price.

Sometimes the logical form is briefer than the original sentence, as in the following example :

A man may be a scholar without being wise.

I. The Subject is simply *scholars.*

II. The Predicate is *wise men.*

III. The relation between the Subject and the Predicate is such that the proposition is Particular and Negative.

So we have a particular negative proposition with *scholars* as subject, and *wise men* as predicate, thus :

Some scholars are not wise men.

Propositions whose subject is an individual, e.g. *Winston Churchill is an Austrian*,[1] are called **singular propositions.** They are in some ways exceptional and require special treatment. If we insist rigidly on maintaining the convention that a proposition states a relation between two classes of entities,[2] then we are forced into treating singular propositions as in the following example :

Churchill is an Austrian

I. Using the same highly conventionalized (and questionable) method, we look for the class of entities about which an assertion is being made. It happens that this is a case where the class has only one member, but this, according to the convention, does not make any relevant difference, so we get as Subject *Churchills* or *persons who are Churchill*.

II. The Predicate is clearly *Austrians*.

III. The relation between the Subject and the Predicate is obviously such that the proposition is Affirmative, and it is also Universal, for the assertion is made about every member of the Subject class, and this is not affected by the further fact that the Subject class consists of only one member.

So we have a universal affirmative proposition with *persons who are Churchill* as subject and *Austrians* as predicate, thus :

All persons who are Churchill are Austrians.

---

[1] The proposition is false, but this is irrelevant to the form of it, which is what we are concerned with here.

[2] Cf. page 5 and note.

This is, of course, a most odd way of stating that Churchill is an Austrian, even though it accords with the convention and has the simplifying advantage of treating classes in the same way whether they have one member or more than one. It is better to write singular propositions straightforwardly, e.g.

Churchill is an Austrian,

bearing in mind that, in spite of the singular verbs and nouns, the relation between subject and predicate is the same as in the previous alternative formulation, i.e. it is affirmative, and it is universal because the predicate applies to the whole of the subject. That is to say, singular propositions are, formally, special cases of universal propositions.

Few philosophers are wealthy.

I. It may at first appear that the Subject is *philosophers,* but consideration shows that what is being commented on is not *philosophers,* but *the proportion of wealthy philosophers,* which is noticed as being small. So the Subject is *the proportion of wealthy philosophers.*

II. The Predicate is *a small proportion.*

III. The relation between the Subject and the Predicate is similar to that of the immediately preceding example, i.e. the proposition is Universal and Affirmative.

So we have a universal affirmative proposition (a singular proposition similar to the previous example) having *the proportion of wealthy philosophers* as subject, and *a small proportion* as predicate, thus:

The proportion of wealthy philosophers is a small proportion.

The tradition is not uniform here, and some logicians would restate the given sentence as *Some philosophers are not wealthy persons*, and would similarly restate *A few X are Y* as *Some X are Y*. This is less clumsy and avoids the use of a class as a collective term in a proposition, but is on the other hand an inaccurate rendering, as the sense of fewness has been lost. As this is a point on which the traditional conventions are inadequate, there are objections to whatever is done about it. A good case can be made for each of the alternatives, and the treatment in the text is chosen as being on the whole the less misleading.

Four men were wounded.

I. This is in some respects similar to the previous example, for the Subject is not *four men* but *the number of men who were wounded*.

II. The Predicate is *the number four*.

III. Again the relation between the Subject and the Predicate is such that the proposition is Universal and Affirmative.

So we have a universal affirmative proposition similar to the two previous examples, having *the number of men who were wounded* as subject, and *the number four* as predicate, thus :

The number of men who were wounded is the number four.

It is conventional in the traditional logic that the copula be in the present tense. Sentences whose verbs are in other tenses thus raise peculiar difficulties if we attempt to express them in logical form. The traditional treatment is to state them so that the significance of pastness or futurity, as the case may be, is contained within the terms. This allows of a copula in the present.

The Hittites were not a Semitic people.

I. The Subject is not *Hittites*, which would entail a verb in the past tense, but *persons who were Hittites*.

II. For similar reasons the Predicate is *persons who were Semites*.

III. The relation between the Subject and the Predicate is clearly such that the proposition is Universal and Negative.

So we have a universal negative proposition with *persons who were Hittites* as subject, and *persons who were Semites* as predicate, thus :

No persons who were Hittites are persons who were Semites.

By similar treatment sentences expressing possibility or probability can be forced into the conventional scheme.

It will probably rain when the barometer is low.

I. The Subject is *occasions on which the barometer is low*.

II. The Predicate is *occasions on which rain is probable*.

III. The relation between the Subject and the Predicate is such that the proposition is Universal and Affirmative.

So we have a universal affirmative proposition with *occasions on which the barometer is low* as subject, and *occasions on which rain is probable* as predicate, thus :

All occasions on which the barometer is low are occasions on which rain is probable.

The traditional logic thus insists upon a high degree of conventionalized simplification in restating the complex expression of ordinary speech and expressing them in one or other of the four recognized forms.

The critical reader will from time to time feel dubious about this treatment of the proposition, but such difficulties must be reserved for later discussion, as we are here concerned only to familiarize ourselves with the long-established conventions that constitute the traditional logic.

As there should by this time be little difficulty in expressing sentences in logical form, we can advance in a new chapter to a further development of doctrine.

NOTE ON NAMES OF TERMS

Logicians have made many attempts to classify terms and give technical names to the principal kinds. The inevitable lack of uniformity in these attempts arises from their being attempts at classifying kinds of meaning rather than kinds of form, and thus raising wide philosophical issues beyond the scope of formal logic proper. However, there is a measure of agreement in the usage of some of these technical names, and the student will find it helpful to be acquainted with the following :

concrete term, abstract term, singular term, general term, collective term.

A term is **concrete**	if it means a thing or a person, e.g. *brick, secretary, typewriter.*
A term is **abstract**	if it means a quality or attribute, e.g. *usefulness, triangularity.*
A term is **singular**	if it means a single entity only, e.g. *Penyghent, Julius Caesar, the richest British subject now alive.*
A term is **general**	if it means any one of an indefinite number of entities, e.g. *book, typewriter, soldier.*
A term is **collective**	if it means a number of entities considered together as one whole, e.g. *regiment, flock, class.*

CHAPTER II

SYMBOLS AND DISTRIBUTION

IN the examples so far examined we have paid no attention to the truth or otherwise of the propositions, but have considered only their form, and it has become progressively clearer that the form of propositions appears to be definite, and to have characteristics of its own which are independent of what is asserted in the proposition, and independent of the truth or otherwise of that assertion. We have found that a proposition having a given subject and a given predicate must, according to the traditional logic, take one or other of only four possible forms. Thus if a proposition has *ungulates* as subject and *carnivores* as predicate, it must be either

All ungulates are carnivores	(universal affirmative)
or *No ungulates are carnivores*	(universal negative)
or *Some ungulates are carnivores*	(particular affirmative)
or *Some ungulates are not carnivores*	(particular negative).

The advance referred to at the close of the last chapter is that of replacing the subject and predicate by symbols, and thereby making it possible to examine the form of the propositions without regard to their specific meaning. If the fully stated subject is replaced by the symbol *S*, and the fully stated predicate is replaced by the symbol *P*, then the four possible forms that a proposition can take are seen to be:

All S are P
No S are P
Some S are P
Some S are not P.

A further step can then be made, replacing both the adjective of quantity and the copula (i.e. the two factors that determine the structure or form of the proposition) by a single symbol.

All S are P   can be symbolized by SaP
No S are P   can be symbolized by SeP
Some S are P   can be symbolized by SiP
Some S are not P   can be symbolized by SoP.

The symbols S and P are, of course, the initial letters of *Subject* and *Predicate*, and the symbols for the relation between them are the vowels of *AffIrmo, I affirm*, and of *nEgO, I deny*.

For convenience and brevity of reference, the universal affirmative proposition, SaP, may be called an 'A proposition'; the universal negative proposition, SeP, may be called an 'E proposition'; the particular affirmative proposition, SiP, may be called an 'I proposition'; and the particular negative proposition, SoP, may be called an 'O proposition'. These symbols must be memorized.

The consequence of this use of symbols is that we can consider SaP by itself as a proposition, without giving specific meanings to S and P, and without raising the question whether the proposition would be true or false if we did give specific meanings to S and P. In other words, there seems to be a *form of proposition*, or *propositional form*, which we can consider by itself, just as in algebra we can consider $(a + b)^2 = a^2 + 2ab + b^2$ without reference to the values of a and b, that is to say without reference to what it is that a and b stand for. The traditional formal logic thus achieves a remarkable simplification, for in examining the structure of arguments, statements, inferences and so forth we have no longer to deal with the innumerably diverse expressions of the English idiom, but only with four comparatively simple forms of propositions ; so simple indeed that we can indicate the mere form by symbols, and can discuss that form without reference to any specific meaning that the symbols might be intended to represent.

On examination each of those forms will be seen to possess, as a mere form, characteristics of its own, but these are difficult to detect without the assistance of another branch of the traditional teaching, namely the doctrine of the distinction between the **denotation** and the **connotation** of a term. This may appear at the moment to have nothing to do with the matter in hand, but its relevance and convenience for this purpose will shortly become clear.

The conventional distinction between the denotation and the connotation of a term may best be explained by taking a simple example. Were I to remark that blue tits are lively little creatures, I might be asked ' What are blue tits ? ' To that I should naturally respond in one or other of two ways. I might take the inquirer out of doors and point at certain birds, saying something like the following : ' That bird clinging upside-down to the top branch is a blue tit, and so is the bird at the far end of the lower branch to the left ' ; or I might instead give a description of blue tits as being small birds between four and five inches long, having olive-green backs, with a bluish tint on the wing and tail feathers, blue and white head markings, and so forth. That is to say, I should explain the meaning of the words I used either by pointing out examples, or by giving a general description.

So when I am asked what blue tits are, i.e. what is the meaning of one of the terms used in the proposition

All blue tits are lively little creatures,

there are two methods of answering.

The one method is to point out the entities, or some of them, to which the term refers, i.e. the entities that the term *denotes*. The class of entities that a term thus denotes is technically called the **denotation** of that term.

The other method of answering is to give a general description by stating the qualities and attributes that the term *connotes*. The qualities and attributes that the term thus connotes are technically called the **connotation** of that term.[1]

It is sometimes said that denotation and connotation vary inversely. The connotation of *M.P.* is *the quality of being a*

---

[1] The introduction of these technical names enables the convention (of treating all propositions as stating a relation between two classes of entities) to be technically expressed as the convention of treating the terms of propositions in denotation (or denotatively). The words *extension* and *intension* are sometimes used in place of *denotation* and *connotation* respectively, e.g. " The traditional logic treats terms in extension (or extensively) ".

3

member of the House of Commons, and the denotation of *M.P.*
is the men and women, about six hundred in number, to whom
that description is applicable. If the connotation is increased
by the addition of the qualification *Conservative*, the term is
then *Conservative M.P.*, and the denotation is at once considerably reduced; and if the connotation is again increased
by making the term *Welsh Conservative M.P.*, then the denotation is smaller still. The assertion that connotation and denotation vary inversely is, however, only approximately true, and
it is mentioned here mainly to familiarize the student further
with the distinction between the denotation and connotation of
a term. The importance of this distinction becomes clear if we
return to the examination of the forms of propositions, namely
SaP, SeP, SiP and SoP, and apply the distinction in
their interpretation, for we can discover a great deal about
the denotation of the terms S and P as there used, even though
we confine ourselves entirely to the forms as forms, without reference to any specific meaning they might be employed to convey.

It can readily be seen that propositions may concern the whole
of the denotation of their terms, or a part of the denotation
of their terms. In *All men are mortal* the assertion is made
about the whole of the denotation of the subject *men*, and in
Some men are moribund the assertion is made about only a part
of the denotation of the subject *men*. The technical words
distributed and **undistributed** are conveniently employed
here, a term being **distributed** in a proposition if an assertion
is made involving the whole of its denotation, and being **undistributed** if an assertion is made involving only a part of its
denotation.

Let us now examine the distribution or otherwise of the
terms in each of the four forms of proposition SaP, SeP,
SiP, and SoP.

THE UNIVERSAL AFFIRMATIVE PROPOSITION
(SaP, The A proposition)

Here we assert that *All* S are P, making an assertion about
the whole of the denotation of the subject. In technical phrase,

the subject of this proposition is said to be distributed, the *S* of *S a P* is distributed.

A diagram will make the position clearer. If we represent the denotation of the subject term by a circle and the denotation of the predicate term by a circle Ⓟ then *S a P*, the proposition that *all S are P*, would be represented either by this diagram :

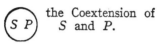

the Inclusion of *S* in *P*.   *All Canadians are British Subjects*

or by this   the Coextension of   *All equilateral tri-*
               S and *P*.   *angles are equi-*
                                       angular triangles,

these two being the only possible relations of *S* and *P* if *All S are P*.

This way of regarding the situation may at first appear confusing, but a little imaginative consideration shows how simple it really is, for the diagram makes clear that in each case the *whole* of the denotation of *S* coincides with at least a part of the denotation of *P*, i.e. it shows that *S* is distributed.

The fact that the subject of the universal affirmative proposition is distributed can be very simply shown symbolically thus, *S̄ a P*, where we merely add the symbol for distribution, viz. –, to the already familiar *S a P*. This is, of course, suggested by the sign for a long vowel in prosody.

As for the predicate, it is to be observed that we assert that *All S are P*, not that *All S are all P*, which might or might not be the case. *All Canadians are British Subjects* does not mean that all Canadians are *all* British Subjects. We speak not of the whole of the denotation of *P*, but only of that part of it that coincides with the denotation of *S*. About the remainder of the denotation of *P*, if there is any remainder, the proposition gives us no information. It happens that in the second example quoted we know that the whole of the denotation of *P* does coincide with the denotation of *S*, but that comes from our knowledge of geometry and not from our knowledge of the form of the proposition. That is, the predicate of the proposition is undistributed, the P of SaP is undistributed.

By using another sign similarly suggested by prosody, this can be represented by $Sa\breve{P}$, which combines with what we have learned of the distribution of the subject to give $\breve{S}a\breve{P}$ to symbolize the universal affirmative form of proposition and the distribution or otherwise of its terms.

THE UNIVERSAL NEGATIVE PROPOSITION

(SeP, The E proposition)

In this proposition we assert that *No S are P*; we again make an assertion about the whole of the denotation of the subject. The S of SeP is in consequence distributed.

In this case there is a very simple diagram:

$\textcircled{S}$ $\textcircled{P}$ the Exclusion of S from P,   *No fish are mammals,*

which is the only possible relation of S and P if *No S are P*. This shows how the *whole* denotation of S is involved in the assertion, i.e. it shows that S is distributed in SeP. This can be represented by $\breve{S}eP$.

As for the predicate, the diagram shows how the proposition *No S are P* asserts the total exclusion of the denotation of S from the whole of the denotation of P, and not merely from part of it. *No fish are mammals* excludes fish from the whole of the class of mammals and not merely from part of it. That is to say, the predicate of the proposition is distributed, the P of SeP is distributed.

Symbolically this is shown by $Se\breve{P}$, which combines with what we have learned of the distribution of the subject to give $\breve{S}e\breve{P}$.

THE PARTICULAR AFFIRMATIVE PROPOSITION

(SiP, The I proposition)

In this proposition we assert only that *Some S are P*; we make an assertion about only a part of the denotation of S.

That is, the subject of this proposition is undistributed, the *S* of *SiP* is undistributed.

Since we make an assertion about only a part of the denotation of *S*, and know nothing further, we neither know which part of *S* is concerned, nor the relation of the remainder of *S*, if any, to *P*. There are in consequence four possibilities to be represented by the diagram, thus:

	the Intersection of *S* and *P*,	*Some poets are novelists*
	the Inclusion of *S* in *P*,	*Some Celts are Aryans* [1]
	the Inclusion of *P* in *S*,	*Some soldiers are gunners*
	the Coextension of *S* and *P*,	*Some rectangles with their adjacent sides equal are squares* [1].

These are all the relations of *S* and *P* that are possible if *Some S are P*. All that these four have in common is that at least a part of the denotation of *S* coincides with at least a part of the denotation of *P*. That is, the *S* of *SiP* is undistributed, which can be represented by *ŠiP*.

As for the predicate, the same considerations show that in this proposition it also is undistributed, for we assert only that *Some S are P*, not that *Some S are all P*. We speak not of the whole of the denotation of *P*, but only of that part of it that is also *S*. About the remainder of the denotation of *P*, if any, the proposition gives no information, as is shown by there being four possible alternatives, any one of which may be the case when *Some S are P*. In other words, the

---

[1] Sentences such as these are not likely to occur in normal discussion, because it is known that *all* Celts are Aryans, and that *all* rectangles with their adjacent sides equal are squares. The sentences are quoted only to exemplify the four possible situations in which *Some S are P.*

predicate of the proposition is undistributed, the P of SiP is undistributed. This can be represented by $Si\breve{P}$, which combines with what we have learned of the distribution of the subject to give $\breve{S}i\breve{P}$.

THE PARTICULAR NEGATIVE PROPOSITION

(SoP, The O proposition)

In this proposition we assert only that *Some S are not P* ; we make an assertion about only a part of the denotation of S. Hence the subject is undistributed, the S of SoP is undistributed.

In this case also we make an assertion about only a part of the denotation of S, and know nothing further. We neither know which part of S is concerned, nor the relation of the remainder of S, if any, to P, so there are alternative possibilities to be represented by the diagram, thus,

the Intersection of S and P, *Some students are not men*

the Inclusion of P in S, *Some game birds are not pheasants*

the Exclusion of S from P, *Some hexagons are not pentagons.*[1]

which are all the possible relations of S and P, if *Some S are not P*. All that these three have in common is that at least a part of the denotation of S does not coincide with the denotation of P or with any part of it. That is, the S of SoP is undistributed, which is represented by $\breve{S}oP$.

As for the predicate, the same examination shows that it is distributed, for part of S is excluded not merely from a part of the denotation of P, but from the whole of the denotation of P. That is, the P of SoP is distributed. This may be represented by $So\bar{P}$, which combines with what we have learned of the distribution of the subject to give $\breve{S}o\bar{P}$.

---

[1] cf. page 29, note.

The distribution of the terms of the four possible forms of proposition can then be summarized symbolically thus,

$$S\, a\, \breve{P}$$
$$S\, e\, \bar{P}$$
$$\breve{S}\, i\, \breve{P}$$
$$\breve{S}\, o\, \bar{P}.$$

Or, otherwise stated,

the subjects of universal propositions are distributed,
the subjects of particular propositions are undistributed,
the predicates of affirmative propositions are undistributed,
the predicates of negative propositions are distributed.

In this chapter we have examined the forms of propositions merely as forms, using throughout symbols only, and we find on the one hand that the forms are comparatively simple, and on the other hand that each form has a number of definite characteristics of its own.

We are now in a position to advance in the next chapter to a consideration of the relations in which a proposition can stand to other propositions.

CHAPTER III

IMMEDIATE INFERENCE

IF it is true that " all barristers are lawyers ", then it is true that " some lawyers are barristers ", and it is false that " some barristers are not lawyers ". This exemplifies the kinds of relation in which propositions can stand to each other, such that if one proposition is true, certain others are true and certain others are false. We habitually make statements dependent for their cogency on the existence of such definite relations between propositions, but we seldom pay attention to those relations as such, and may even be slightly surprised to discover that they exist at all. Those relations are now to be systematically examined, the different kinds being distinguished and their characteristics noted.

We shall find that the characteristics of these relations can be discovered by examining the form of the propositions, and indeed it is only by considering the form alone, that is by using the symbolic representation, that we can adequately understand those relationships and can be certain not to overlook any of the possible varieties. We shall therefore pay primary attention to the symbols representing the form of the propositions, referring to propositions fully expressed in words merely by way of illustration from time to time.

Let us begin by considering as an example the universal affirmative proposition and inquiring what other propositions having the same terms are true if it is true. If it is true that *All ungulates are carnivores*, then it is also true that *Some ungulates are carnivores*, i.e. if *All S are P* then *Some S are P*. This latter may seem a meagre item of information, of lesser importance than the original proposition, but even so, it is true if the original proposition is true. That is, the two propositions are so related that if SaP is true, then SiP is true also.

This has brought us to the topic of **inference**, for we have found certain propositions to be so related that if one of them is true, then the other is true also, which is to say that the latter can be **inferred** from the former.[1] By discovering the different relations in which one proposition can stand to other propositions having the same terms, we at the same time discover what inferences can validly be made from one proposition to other propositions having the same terms. By examining these different kinds of relation we examine at the same time the different kinds of immediate inference.[2]

In this examination it is illuminating to observe the distribution of the terms. In the simple example quoted, where *Some ungulates are carnivores* is inferred from *All ungulates are carnivores*, the original proposition is $S a \breve{P}$ and the derivative is $\breve{S} i \breve{P}$. No increase in distribution has been made in either of the terms, and indeed the subject which was given distributed is undistributed in the derivative. We were given information covering the whole of the denotation of S, and we are employing information about a part of it only, and that is why *Some ungulates are carnivores* seems a meagre statement compared with the universal proposition from which it is derived. But, meagre or not, it is true if the original proposition is true, that is to say,

---

[1] The former, being given or premised, is conveniently called the **premiss** and the latter, being inferred or concluded, is conveniently called the **conclusion**. **Immediate inference,** the name of this very simple kind of inference, does not of course mean that the conclusion is drawn ' without delay in time ', but that it is drawn ' without the intervention or mediation of any other factor '.

[2] At this point the distinction between truth and validity, which we employ somewhat casually and even loosely in everyday thinking, becomes essentially important and must be clearly understood. **Only propositions can be true or false, and only inferences can be valid or invalid.** A proposition is true if it corresponds to the facts, while an inference is valid if its conclusion follows from its premiss, whether that premiss is itself true or false. (This naïve and question-begging definition of truth and validity is intended only to make the distinction clear. Any further discussion would be out of place here, for the question involves a man's whole philosophy.) A valid inference is valid whether the premiss is true or false, and an invalid inference is invalid whether the premiss is true or false. It is also, of course, to be borne in mind that a conclusion validly inferred is not *on that account* true (since the premiss might have been false), and a conclusion invalidly inferred is not *on that account* false (since it might happen to be true for other reasons).

the inference from the given universal proposition to the particular is valid.

The doctrine of the distribution of terms enjoys some practical importance here, for a brief consideration shows clearly that **if an inference is to be valid, terms that are undistributed in the original proposition must remain undistributed in the derivative proposition.** This principle is quite common-sense and straightforward, and with it in mind we can readily distinguish the different types of inference, and understand why some are valid and why others would be invalid. Let us now examine systematically the different kinds of relation between propositions having the same terms, beginning with the simple kind that has already served as an example.

SUBALTERN RELATION [1]

We have already seen that $S a P$ and $S i P$ are so related that if $S a P$ is true, then $S i P$ is true also, i.e. from $S a P$ we can infer $S i P$. This relation (the relation in which a particular proposition stands to the universal proposition of the same quality) is called the **subaltern** relation, $S i \breve{P}$ being the subaltern of $S a \breve{P}$. For exactly similar reasons, $S o \breve{P}$ is subaltern to $S e \breve{P}$, i.e. from $S e \breve{P}$ we can infer $S o \breve{P}$.

To spend time on inferences so uninteresting may seem disproportionate, but the importance of this simple kind of immediate inference and its connexion with the other kinds will later become intelligible if we describe what we have been doing as discovering what other propositions having S as subject and P as predicate can be validly inferred from a given proposition having S as subject and P as predicate.

This is the first of the five noteworthy types of relation in which propositions having the same terms can stand to each other.

It is to be noticed that this relation is not reciprocal, for if we are given the particular proposition that *Some S are P*, we are not entitled to infer that *All S are P*. That might or might not in fact be true, but whether so or not, we cannot *infer* it. The symbolic treatment shows this clearly, for the

---

[1] Occasional reference to the schematic summary on page 98 may be found helpful from this point onward.

original proposition is $S i \breve{P}$, and the alleged inference would be $S a \breve{P}$, with S distributed, which would be invalid, as S in the original proposition was undistributed. We are given a statement about a part of the denotation of S, and we are in consequence not entitled to make any inference involving the whole of the denotation of S.

The universal $S a \breve{P}$ and its subaltern $S i \breve{P}$ are so related to each other that if $S a \breve{P}$ is true then we can validly infer that $S i \breve{P}$ is true also, but if $S i \breve{P}$ is true, we are not entitled to infer *therefrom* that $S a \breve{P}$ is true.[1]

For exactly similar reasons we can infer $S o \breve{P}$ from $S e \breve{P}$, but cannot infer $S e \breve{P}$ from $S o \breve{P}$.

In tabular form,

> From $S a \breve{P}$ we can infer $S i \breve{P}$.
>
> From $S i \breve{P}$ we cannot infer any proposition with S distributed.
>
> From $S e \breve{P}$ we can infer $S o \breve{P}$.
>
> From $S o \breve{P}$ we cannot infer any proposition with S distributed.

CONVERSE RELATION

If it is true that no lawyers are honest men, we can infer that no honest men are lawyers. In logical phraseology, what we have done in this kind of immediate inference is to infer from a proposition with S as subject and P as predicate another proposition with P as subject and S as predicate. This form of immediate inference is called **conversion,** and the P-S proposition is called the **converse** of the S-P proposition, the two propositions standing to each other in converse relation. We shall examine the four forms of propositions and their converses, if any, paying special attention to the distribution of the terms.

CONVERSE OF AN E PROPOSITION

The symbolic form of the example quoted, *No lawyers are honest men,* is, of course, $S e \breve{P}$, and that of its converse *No honest men are lawyers* is $\breve{P} e S$. Both terms of the new proposition are distributed, but they were distributed in the original proposition.

[1] Cf. page 33, note 2.

The original proposition made an assertion about the whole of the denotation of S and the whole of the denotation of P, and we are consequently entitled to make the new assertion about the whole of the denotation of S and the whole of the denotation of P.

The diagram, representing the proposition *No S are P*, makes this clear,

(S) (P) *No fish are mammals*,

for the same diagram also represents the proposition *No P are S*; i.e. if $Se\bar{P}$ is true, then $\bar{P}e\bar{S}$ is true also, i.e. from $Se\bar{P}$ we can infer $\bar{P}e\bar{S}$.

For exactly similar reasons we can infer $Se\bar{P}$ from $\bar{P}e\bar{S}$. In other words, the relation of an E proposition to its converse is reciprocal.

CONVERSE OF AN I PROPOSITION

The conversion of a particular affirmative proposition also is quite straightforward. From the statement that *Some sailors are gentlemen* we can infer that *Some gentlemen are sailors*. In symbols, if we are given $Si\bar{P}$, we can infer $\bar{P}iS$. In this case both S and P are undistributed in the original proposition, but they are undistributed in the derivative proposition also, and the inference is in consequence valid.

Again the diagram representing the proposition *Some S are P* is helpful,

$(S\!\!\!(P))$ *Some poets are novelists*

$((S)P)$ *Some Celts are Aryans*

$(S(P))$ *Some soldiers are gunners*

$(S\ P)$ *Some rectangles with their adjacent sides equal are squares*, for each of these diagrams also represents the proposition *Some P are S*. I.e. if $S i \breve{P}$ is true, then $\breve{P} i S$ is true also, i.e. from $S i \breve{P}$ we can infer $\breve{P} i S$.

For exactly similar reasons we can infer $S i \breve{P}$ from $\breve{P} i S$. In other words, the relation of an **I** proposition to its converse is reciprocal.

CONVERSE OF AN O PROPOSITION

The case of the particular negative proposition is not so simple. If we know only that *Some roans are not bays*, we cannot tell anything about bays. For all we know, *all* bays might be roans, or *none* might be, or *some* might be, or *some* might not. These four alternatives are all compatible with its being the case that *Some roans are not bays*. Therefore there is no *P-S* proposition that can be inferred from an *S o P* proposition. That is, the particular negative proposition has no converse. The diagram shows this, for if we know only that *Some roans are not bays* then the situation might be :

 in which case *Some bays are roans, P i S*, would be true
and *Some bays are not roans, P o S*, would be true,

or it might be :

 in which case *All bays are roans, P a S*, would be true,

or it might be :

 in which case *No bays are roans, P e S*, would be true.

From the information given in *S o P* we cannot tell which of the four *P-S* propositions is true. In other words, from *S o P* no valid converse can be inferred.[1]

It is possible to explain more simply the inconvertibility of an **O** proposition, by examining the distribution of its terms.

[1] Cf. page 33, note 2.

The attempt to convert $\breve{S}o\bar{P}$ would make S serve in a new capacity as predicate of a negative proposition. The predicates of negative propositions are distributed, and therefore S would be distributed in its new capacity. But it is undistributed in the original proposition, $\breve{S}oP$, and therefore the alleged conversion would be invalid.

CONVERSE OF AN A PROPOSITION

The conversion of the universal affirmative proposition is interesting, and because of its peculiarity has received much attention from the logicians. If I say *Thirty days hath September*, which in logical form is *All Septembers are periods of thirty days*, I am not entitled to infer that *All periods of thirty days are Septembers*, which is plainly not true, but only that *Some periods of thirty days are Septembers*. A reference to the distribution of the terms shows this clearly, for the original proposition is $Sa\breve{P}$, and P must consequently remain undistributed in its new position as subject of the converse. Only particular propositions have an undistributed subject, so the converse must be a particular proposition, i.e. it must be $\breve{P}iS$. If we attempt to convert $Sa\breve{P}$ to a universal proposition we should get $\bar{P}aS$, which uses P distributed and is hence invalid. The latter procedure is a fallacy not unknown in discussion or controversy, e.g. if it is agreed that *All persons who cannot control themselves are morally unsatisfactory* it is sometimes alleged as an inference therefrom that *All morally unsatisfactory persons are persons who cannot control themselves*, which is, of course, invalid.[1]

These results may be tabulated thus:

the converse of $Sa\breve{P}$ is $\breve{P}i\breve{S}$
the converse of $Se\bar{P}$ is $\bar{P}e\breve{S}$
the converse of $Si\breve{P}$ is $\breve{P}i\breve{S}$
and $\breve{S}o\bar{P}$ has no converse.

[1] The conversion of SaP to PiS is sometimes called conversion *per accidens*, and the illegitimate conversion of SaP to PaS is sometimes called the Fallacy of Simple Conversion of an A proposition.

The converse relation is the second of the types of relation of propositions that must be known and understood.

OBVERSE RELATION

The third of these types is the **obverse relation,** giving rise to the somewhat artificial kind of inference called **obversion.** To understand obversion we must first understand what is signified by the new symbols *not-S* and *not-P*. If the symbol *P* in a proposition stands for *men*, then it is reasonable enough to use a symbol such as *not-P* to stand for what is *not* a man, to stand for anything and everything *other than men*. Whatever *P* may represent, *not-P* represents everything else in the universe.

This must be clearly understood, for the loose interpretation of *not-P* causes many mistakes. If *P* means *men*, then *not-P* does not mean *women and children*; it means everything in the universe other than men, and hence, of course, includes women and children among the host of other entities, things, events, acts, thoughts and so forth that are *other than men*.

From a proposition with *S* as subject and *P* as predicate we can infer another proposition with *S* as subject and *not-P* as predicate. From the proposition *All nurses are women* we can infer its obverse, namely *No nurses are beings other than women*. The symbolic representation of this process of obversion is that we are given *S a P*, and that from it we infer *S e not-P*. The obverses of the other three forms of propositions are equally simple, thus:

the obverse of *S e P* is *S a not-P*
the obverse of *S i P* is *S o not-P*
the obverse of *S o P* is *S i not-P*.[1]

[1] The importance of obversion is small, and it has a place in the traditional logic mainly for the sake of completeness, and because it may be employed in conjunction with conversion to evolve immediate inferences of a high degree of complexity. If, for example, we obvert a proposition and then convert the new proposition (i.e. convert the obverse of the original proposition), we obtain a proposition that is sometimes called the contrapositive of the original proposition. Various combinations of conversion and obversion produce other and even more complicated developments, but all these artificial elaborations can safely be neglected, provided the student is familiar with the three simple relations, namely the subaltern, converse and obverse.

This completes our examination of what propositions are *true* if a given proposition is true. Let us now consider what propositions are *false* if a given proposition is true. There are two kinds of relation (the **contrary** and the **contradictory**) in which propositions can stand to each other such that if one is true the other is false. These two join the three already examined to complete the five noteworthy types of relation in which propositions having the same terms can stand to each other.

CONTRARY RELATION

If *All Clydesdales are Suffolk Punches* were true, then *No Clydesdales are Suffolk Punches* would be false ; and if *No Clydesdales are Suffolk Punches* were true, then *All Clydesdales are Suffolk Punches* would be false.

> I.e. if $S a P$ is true, then $S e P$ is false
> if $S e P$ is true, then $S a P$ is false.

The two universal propositions of opposite quality (having, of course, the same subject terms and predicate terms) are said to be in contrary relation, each being the contrary of the other. If one of them is true, then the other is false.

CONTRADICTORY RELATION

This is the last, and most important, of the five noteworthy types of relation between propositions.

If we wish to contradict the assertion that *All S are P,* we are not required to show that *No S are P* ; it is sufficient if we show that *Some S are not P.* If I object to an assertion that all politicians are self-seekers and wish to refute it, I can do so without having to show that no politicians at all are self-seekers. It is sufficient if I produce *some* politicians who are not self-seekers, even only one, in order to refute the assertion that all politicians are self-seekers. In symbols, $S a P$ is contradicted by $S o P$. ($S e P$ also, of course, contradicts $S a P$, but it gives more information than the minimum necessary to contradict $S a P$.)

On the other hand, if I wish to contradict the assertion that *Some S are not P,* a particular proposition is not sufficient, and the universal proposition *All S are P* is required. If I wish to contradict the statement that some newspapers are not reliable, I must do more than show that some newspapers are reliable, for I must go further and show that *all* newspapers are reliable.

That is to say, SaP and SoP stand in contradictory relation. Each is both necessary and sufficient to contradict the other.

Similarly SeP and SiP stand in contradictory relation, for if I wish to contradict the assertion that *No S are P* it is both necessary and sufficient to show that *Some S are P.* And likewise, if I wish to contradict the assertion that *Some S are P*, I have to show that *No S are P.*

In other words, each universal proposition and the particular of the opposite quality are said to stand in contradictory relation, each being the contradictory of the other. This has to be distinguished carefully from the contrary relationship which holds between universal propositions only.

These five kinds of relation can be exemplified as follows :

Given the sentence *Men are deceivers ever*, which in logical form is *All men are habitual deceivers*, $S a \bar{P}$, a universal affirmative proposition, then

the Subaltern	is	*Some men are habitual deceivers*
the Converse	is	*Some habitual deceivers are men*
the Obverse	is	*No men are beings other than habitual deceivers*
the Contrary	**is**	*No men are habitual deceivers*
the Contradictory	is	*Some men are not habitual deceivers*

Given the sentence *Oils won't mix with water*, which in logical form is *No oils are substances that mix with water*, $S e \bar{P}$, a universal negative proposition, then

the Subaltern	is	*Some oils are not substances that mix with water*

4

the Converse	is	*No substances that mix with water are oils*
the Obverse	is	*All oils are entities other than substances that mix with water*
the Contrary	is	*All oils are substances that mix with water*
the Contradictory is		*Some oils are substances that mix with water.*

Given the sentence *Some biologists think he is right*, which in logical form is *Some biologists are persons who think that he is right*, SiP, a particular affirmative proposition, then

the Subaltern		(There is no Subaltern)
the Converse	is	*Some persons who think that he is right are biologists*
the Obverse	is	*Some biologists are not entities other than persons who think that he is right*
the Contrary		(There is no Contrary)
the Contradictory is		*No biologists are persons who think that he is right.*

Given the sentence *Scholars aren't all of them wise*, which in logical form is *Some scholars are not wise men*, SoP, a particular negative proposition, then

the Subaltern		(There is no Subaltern)
the Converse		(There is no Converse)
the Obverse	is	*Some scholars are beings other than wise men*
the Contrary		(There is no Contrary)
the Contradictory is		*All scholars are wise men.*

The system of the simple relationships in which propositions (propositions having, of course, the same terms) can stand to each other can be tabulated symbolically thus :

Given Proposi-tion	Subal-tern	Con-verse	Obverse	Con-trary	Contra-dictory
SaP	SiP	PiS	$S\ e\ not\text{-}P$	SeP	SoP
SeP	SoP	PeS	$S\ a\ not\text{-}P$	SaP	SiP
SiP	—	PiS	$S\ o\ not\text{-}P$	—	SeP
SoP	—	—	$S\ i\ not\text{-}P$	—	SaP

The various possibilities of valid immediate inference are clearly shown here. If, for instance, SeP is true, then we know that PeS, $S\ a\ not\text{-}P$ and SoP are true also, and that SiP and SaP on the other hand are false.[1]

By this time the student should be in a position to understand the traditional view of the structure of propositions, and of the relations in which propositions having the same terms can stand to each other, with the consequent possible kinds of valid immediate inference.

[It is said by some logicians that all reasoning must proceed in accordance with the so-called 'Laws of Thought'. This is debatable, some philosophers holding the epistemological presuppositions on which these 'Laws' are based to be mistaken. Whether this is so or not, however, the 'Laws of Thought' ought to be known to the student. They are three in number.

1. The Law of Identity   (A thing is what it is.)
2. The Law of Contradiction   (A thing cannot both be so-and-so and not be so-and-so.)
3. The Law of Excluded Middle   (A thing must either be so-and-so or not be so-and-so.)

In some formulations the qualification 'at the same time' is added.]

---

[1] We have here dealt with the different kinds of relation between propositions along with the different kinds of immediate inference that depend on these relations.

The older sequence of teaching was to deal with the kinds of relation or 'opposition' of propositions first (treating immediate inferences as a separate and later topic), and a diagram, called the Square of Opposition, was long used to schematize some of these relations or oppositions in an easily remembered form. It is self-explanatory. See page 99.

CHAPTER IV

MEDIATE INFERENCE. SYLLOGISM

THE traditional logic maintains the convention that all statements and arguments, no matter how lengthy or involved, can be analysed and shown to consist of simple propositions, or of groups of simple propositions standing in some systematic relation to each other. In the earlier chapters we dealt first with propositions, and then with that very simple kind of systematic relation of propositions that enables immediate inferences to be drawn. Only a few, however, of the inferences drawn in conversation and in scientific inquiry are of that simple kind, and in most of them we can recognize at sight a higher degree of complexity. We now advance to an examination of the systematic relations in which propositions stand to each other when playing their parts in more complex arguments.

There are two ways of approaching this topic, for we can begin with propositions as units and build up complex arguments, or we can begin with complex arguments and analyse them to discover their constituent propositions. The former is suggested by the manner in which we have so far dealt with formal logic, working from the more simple to the more complex, but we shall at this point take the second alternative as it is easier to follow, for the choice is one of pedagogic effectiveness only.

Consider, with a view to analysing them, examples of complex arguments such as the following :

Thebans are Boeotians and Boeotians are Aeolians, therefore Thebans are Aeolians.

Snuff is a kind of tobacco, and there is a heavy tax on tobacco, so there must be a tax on snuff too.

Anaesthetists are bound by the Hippocratic oath, for all members of the medical profession are, and anaesthetists are members of that profession.

In each of these examples we appear to be given not one statement but two, and from these two taken together we derive a new item of information that we could not derive from either of them taken by itself. This is made very clear by an old anecdote often quoted in this connexion.

While talking of his early experiences as a priest, an elderly abbé responded to the comment that the secrets of the confessional must often be of a kind disturbing to a young man, by admitting that it had indeed been so in his case, as the first confession he ever heard was a confession of murder. Shortly after his departure his visit was mentioned to a later caller, a local proprietor and notability, who remarked that the abbé and he were very old acquaintances. " Indeed," he added, " I was the abbé's first penitent."

In this case, the two items of information were given by different persons, at different times, each unaware of the other's statement, yet the two taken together provided a third item of information that was altogether new.

Many inferences appear on examination to be of this form, for they can be shown to consist of two statements that are given, and of a third statement that is inferred from the two that are given. This kind of argument is typified by the example that was used by Aristotle and by all logicians since,

> *All men are mortal beings*
> *Socrates is a man*
>
> ---
>
> ∴ *Socrates is a mortal being.*

Here the three statements have been expressed in logical form, clearly showing that two propositions are given or premised (the **premisses**), and that a third is inferred or concluded (the **conclusion**), thus exemplifying what is meant by saying that arguments can be expressed as " groups of propositions standing in some systematic relation to each other ". This kind of systematic relation of propositions is called a **syllogism**, συλλογισμός from συλλογίζομαι, to consider together.

It is easy to discover that a syllogism has a form that can

be studied independently of its meaning, just as a proposition has a form that can be studied independently of its meaning. Take an example that we can already recognize to be a syllogism though we may as yet have only a vague notion of its structure :

> *All pedunculated cirripedes are crustaceans*
> *All barnacles are pedunculated cirripedes*
>
> ∴ *All barnacles are crustaceans.*

The important and perhaps surprising point emerges that we must accept the conclusion if we accept the premisses, whether or not we know what pedunculated cirripedes are. Even though we may never before have heard of pedunculated cirripedes we see at once that this conclusion follows inevitably from these premisses. We can therefore replace *pedunculated cirripedes* by a symbol, say *M*, making the syllogism as follows :

> *All M are crustaceans*
> *All barnacles are M*
>
> ∴ *All barnacles are crustaceans.*

If we now substitute symbols for the remaining terms also, replacing the subject of the conclusion by *S* and the predicate of the conclusion by *P*, then we have :

> *All M are P*
> *All S are M*
>
> ∴ *All S are P.*

This can be symbolized throughout, thus :

> *M a P*
> *S a M*
>
> ∴ *S a P,*

and we must always accept this conclusion if we accept these premisses, whatever S and P and M may stand for.

This further discovery is impressive. Not only does there appear to be a *form of proposition*, but there appears also to be a *form of argument* in which the conclusion must be true if the premisses are true, independent of what the terms employed may stand for.

The form we have so far discussed, viz.,

$$M\,a\,P$$
$$S\,a\,M$$
$$\overline{}$$
$$\therefore S\,a\,P$$

is not the only possible form of syllogism, for a differently arranged argument such as

All Platonists are mathematicians
Some Platonists are mystics

∴ *Some mystics are mathematicians*

All M are P	$M\,a\,P$
Some M are S	$M\,i\,S$
∴ *Some S are P*	$\therefore S\,i\,P$

is a syllogism also, though a different kind of syllogism. There are many such kinds, but it is possible to investigate the structure of syllogisms without cataloguing all the variants.

In any syllogism, whatever kind of syllogism it be, one of the premisses must give some information about the subject of the conclusion, and the other premiss must give some information about the predicate of the conclusion. One premiss must hence contain the subject of the conclusion, and, of course, one other term, while the other premiss must contain the predicate of the conclusion and one other term. And that 'other term' must be the same in both premisses, otherwise they would have no connexion with each other.

So in a syllogism three terms are involved, each of them appearing twice, viz.

> *the term that appears as subject in the conclusion,*
> *the term that appears as predicate in the conclusion,*
> *and the " other term " called the middle term.*

This **middle term** does not appear in the conclusion, but the conclusion can be reached only through its mediation. That is why this type of inference is called mediate inference, in distinction from immediate inference in which no mediating or middle term is required.[1]

A syllogism thus consists of a conclusion which is a proposition with S as subject and P as predicate ; and of two premisses, of which one is a proposition having S and M as its terms, and the other a proposition having P and M as its terms. The subject of the conclusion is commonly called the **minor term,** and the premiss in which it appears the **minor premiss.** Similarly the predicate of the conclusion is called the **major term,** and the premiss in which it appears is called the **major premiss.**

The term S is subject in the conclusion, but it may be either subject or predicate in the premiss in which it occurs, (M being, of course, the other term). Similarly P, which is predicate in the conclusion, may be either subject or predicate in the premiss in which it occurs. This may be expressed in other words by saying that M may be either subject or predicate in the first or major premiss, and it may be either subject or predicate in the second or minor premiss, which gives alternative arrangements of S, P and M within the syllogistic form. Examples make this clear.

---

[1] Reference to page 100 will be useful from this point onward.

I. *All men are mortal beings*   M P
    Socrates is a man   S M

∴ *Socrates is a mortal being*   ∴ S P

II. *All judges are lawyers*   P M
    No bishops are lawyers   S M

∴ *No bishops are judges*   ∴ S P

III. *All Platonists are mathematicians*   M P
    Some Platonists are mystics   M S

∴ *Some mystics are mathematicians*   ∴ S P

IV. *No Greeks are Trojans*   P M
    Some Trojans are heroes   M S

∴ *Some heroes are not Greeks*   ∴ S P.

These, giving the only possible positions of *S*, *P* and *M*, are known as the four **figures** of the syllogism. Every syllogistic argument will be found to take one or other of these four forms.[1]

An argument may be described as being of a specified figure. Thus the following,

> *Some Zulus are reliable men*
> *All Zulus are Bantus*

> ∴ *Some Bantus are reliable men*

is an argument " in the third figure ".

[1] The fourth figure is uncommon, but arguments occur in the first three figures about equally frequently. To prevent confusing the first and fourth figures, it should be remembered that it is conventional to write the major premiss (*i.e.* that involving *P*) in the top line.

The three propositions constituting a syllogistic argument have their terms arranged in one or other of only four possible ways (i.e. in one or other of the four figures), but in any one of these ways or figures the constituent propositions may be either universal or particular, affirmative or negative. Thus in each of the four figures there are many possible kinds of syllogism according to the quantity and quality of the constituent propositions. These kinds are known as the various modes or **moods** of the figure in question, thus

All men are mortal beings	$M\,a\,P$
Socrates is a man	$S\,a\,M$
∴ *Socrates is a mortal being*	∴ $S\,a\,P$

is a first figure argument, and it is the mood of that figure that has an A proposition as its major premiss, an A proposition as its minor premiss, and an A proposition as its conclusion. It is hence conveniently called the Mood $A\,A\,A$ in the First Figure.

Similarly the argument

Some Zulus are reliable men	$M\,i\,P$
All Zulus are Bantus	$M\,a\,S$
∴ *Some Bantus are reliable men*	∴ $S\,i\,P$

is the Mood $I\,A\,I$ in the Third Figure.

As there are four figures, and in each of them a large number of possible moods, there is in consequence a large but definite number of possible forms,[1] and every syllogistic argument must take one or other of these forms, whatever it be an argument about.

---

[1] There are four alternatives, A, E, I or O in the first Premiss, four similarly in the other Premiss, and four similarly in the Conclusion, giving $4 \times 4 \times 4 = 64$ in each Figure, and there are four Figures, hence the total number of possible variants is 256. Of course, only a very few of them are valid, only 19 in fact.

Only a few of these forms are valid. The following argument for instance, which is *A E E* in the First Figure, is obviously invalid.

All *policemen are civil servants*	*M a P*
No postmen are policemen	*S e M*
∴ *No postmen are civil servants*	∴ *S e P*

It would be possible to go over each of the two hundred and fifty-six variants, finding out by inspection which are valid and which are invalid, but the valid and the invalid can be discriminated by considering the principles on which this kind of reasoning appears to be based, and formulating what are called **Rules of Syllogism,** i.e. rules that must be respected if the syllogism in question is to be a valid argument. The early logicians drew up many such rules, and any student today can with some industry do the same for himself, since these rules are only statements of some of the conditions which must be fulfilled if a syllogism is to be valid. They are implied in what has already been said. Of the various rules which can thus be formulated, there are three which merit special attention and which must be memorized.

1. No term may be distributed in the conclusion if it is not distributed in the premiss in which it occurs.
2. The middle term must be distributed once at least.
3. At least one premiss must be affirmative.[1]

[1] In most text-books there are many other Rules of Syllogism, but it seems hardly necessary to formulate them as rules and remember them as such, since they do no more than state a selection of the many conditions of validity which are sufficiently obvious to common sense. For instance, we may if we wish call it a rule that a syllogism must contain three and only three terms, but this is only another way of stating part of the definition of a syllogism ; similarly we may if we wish call it a rule that if one premiss is negative the conclusion is negative, but this is only another way of pointing out that if one premiss is negative we have excluded one of the terms from the middle term, and that this prevents our deducing anything about the relation of *S* and *P* except that they are in some way excluded from each other, i.e. that the conclusion must be negative. The principal reason for paying special attention to the three rules mentioned

The first of these is easy to appreciate, for if a term is undistributed in its premiss, then that premiss gives information about only a part of the denotation of that term, and there is in consequence no justification for drawing a conclusion about the whole of its denotation, which is what would happen if the term were distributed in the conclusion.

In the patently invalid example quoted,

All policemen are civil servants	$\breve{M} a \breve{P}$
No postmen are policemen	$\breve{S} e \breve{M}$
∴ No postmen are civil servants	∴ $\breve{S} e \breve{P}$

the invalidity arises from the fact that the term *civil servants* is distributed in the conclusion though undistributed in its premiss, i.e. the conclusion makes an assertion about *all* civil servants, while the premiss makes an assertion about *some* of them only, namely about those of them who are policemen.

The term in question is said to have suffered an **illicit process,** and the breach of this rule is known as the **fallacy of illicit process of the major term** or of the **minor term,** as the case may be.

The ground of the second Rule is equally simple, though perhaps less obvious than that of the first Rule, and it can most easily be made apparent by considering a clear example of this fallacy in a third figure argument.

Some students are men	$\breve{M} i \breve{P}$
Some students are women	$\breve{M} i \breve{S}$
∴ Some women are men	∴ $\breve{S} i \breve{P}$

is that they deal with the only kinds of formal mistake in reasoning that are at all likely to pass undetected. The other rules are generally given as follows :

A syllogism must contain three and only three terms.

If one premiss is negative the conclusion must be negative, and vice versa.

If both premisses are affirmative the conclusion must be affirmative, and vice versa.

One of the premisses must be universal.

If one premiss is particular the conclusion must be particular.

As the major premiss is a particular proposition, its subject (the middle term) is not distributed, hence in it we speak of only a part of the denotation of the middle term. As the minor premiss also is a particular proposition, its subject (the middle term) is not distributed, hence we again speak of only a part of the denotation of that middle term, and not necessarily of the same part that was referred to in the major premiss. We cannot in consequence be sure that the two premisses have anything in common, and therefore no inference can be drawn. To ensure that the two premisses do have something in common, i.e. to ensure that the middle term in one of the premisses either includes or is the same as the middle term in the other premiss, that middle term must be distributed once at least. It can readily be seen that this must hold in all the figures of the syllogism. The breach of this Rule is known as the **fallacy of undistributed middle.**

The third Rule (that at least one premiss must be affirmative, i.e. that no conclusion can validly be drawn from two negative premisses) holds because it is only another way of pointing out that if the two premisses are negative we exclude S from M and also exclude P from M, which prevents us from deducing anything about the relation in which S stands to P.

The reader has now sufficient acquaintance with the doctrine of the syllogism to express arguments as syllogisms, and to examine their structure and formal validity. Arguments as colloquially stated seldom look like syllogisms, and in order to express them as syllogisms, drastic restatement may be necessary ; **while terms or even whole premisses which are not stated explicitly but are meant to be " understood " may have to be written out in full.** It will be found simplest to deal with such arguments in the following five stages :

I. Find the Conclusion and express it in Logical Form.[1]

II. State in Logical Form the Premiss that has as its Terms the Middle Term and the Predicate of the Conclusion. (This is the Major Premiss. Cf. page 48.)

III. State in Logical Form the Premiss that has as its Terms

---

[1] This apparently inverted method of restating an argument by beginning at the end is intended to prevent misunderstanding of difficult cases.

the Middle Term and the Subject of the Conclusion. (This is the Minor Premiss. Cf. page 48.)

IV. Write down the syllogism in the conventional lay-out, i.e. with the Major Premiss in the top line, the Minor Premiss in the second line, and the Conclusion in the third line.

V. Inquire (*a*) into the structure of the syllogism and (*b*) into its validity or otherwise, particularly examining to this end the distribution of the Terms, and noting whether at least one Premiss is affirmative.

EXAMPLES OF THE ABOVE

Thebans are Boeotians, and Boeotians are Aeolians, therefore Thebans are Aeolians.

I. The Conclusion is *All Thebans are Aeolians*.

II. The Major Premiss is *All Boeotians are Aeolians*, i.e. the proposition having as its Terms the Middle Term (which is *Boeotians*) and the Predicate of the Conclusion (which is *Aeolians*).

III. The Minor Premiss is *All Thebans are Boeotians*, i.e. the proposition having as its Terms the Middle Term (which is *Boeotians*) and the Subject of the Conclusion (which is *Thebans*).

IV. Written in the conventional way and symbolized, the syllogism is as follows :

$$All\ Boeotians\ are\ Aeolians$$
$$All\ Thebans\ are\ Boeotians$$
$$\therefore\ All\ Thebans\ are\ Aeolians$$

V. (*a*) Structure

$\bar{M}\ a\ \breve{P}$

$\bar{S}\ a\ \bar{M}$

——

$\therefore\ \bar{S}\ a\ \breve{P}$

Mood *A A A* in the First Figure.

(*b*) Validity

The Middle Term has been distributed at least once ; and *S*, which is the only Term distributed in the Conclusion, is distributed in its Premiss. Both Premisses are affirmative. The syllogism is valid.

Indirect taxation is bad, for the tax on newspapers is indirect and it is bad.

I.   The intended Conclusion is *All* (*indirect taxes*) *are* (*bad taxes*).

II.   The Major Premiss is *All* (*taxes on newspapers*) *are* (*bad taxes*).

III. The Minor Premiss is *All* (*taxes on newspapers*) *are* (*indirect taxes*).

IV.   *All taxes on newspapers are bad taxes*
      All taxes on newspapers are indirect taxes
      _____
      ∴ *All indirect taxes are bad taxes*

V.   (*a*) *Structure*

$\bar{M} \, a \, \breve{P}$
$\bar{M} \, a \, \breve{S}$
————   Mood *A A A* in the Third Figure.
∴ $S \, a \, \breve{P}$

(*b*) Validity

The Middle Term has been distributed at least once (it has indeed been distributed in both cases) ; but *S*, which is distributed in the Conclusion, is not distributed in its Premiss, namely the Minor Premiss. The syllogism is therefore invalid through its Illicit Process of the Minor.[1]

Some of the things alleged by the spiritualists are incredible, because they contradict the laws of nature.

I.   The Conclusion is *Some* (*things that the spiritualists allege*) *are* (*incredible things*).

II.   The Major Premiss, which in this example is understood though not explicit in the original statement, is *All* (*things that contradict the laws of nature*) *are* (*incredible things*).

[1] This is one of the syllogistic forms of ' arguing from an example '. Had we been content to draw a particular conclusion, namely *Some indirect taxes are bad taxes*, *S* would have been undistributed in the conclusion, and the syllogism would have been valid.

III. The Minor Premiss is *Some (things that the spiritualists allege) are (things that contradict the laws of nature).*

IV. *All things that contradict the laws of nature are incredible things*
Some things that the spiritualists allege are things that contradict the laws of nature

∴ *Some things that the spiritualists allege are incredible things*

V. (a) Structure

$\bar{M}$ a $\breve{P}$
$\breve{S}$ i $\bar{M}$
─────
∴ $\breve{S}$ i $\breve{P}$

Mood *A I I* in the First Figure.

(b) Validity

M is distributed once; and neither *S* nor *P* is distributed in the Conclusion. Both Premisses are affirmative. The syllogism is valid.[1]

───────────

Bishops who are not yet sufficiently senior to have a seat in the Lords cannot stand for election to the Commons, for no Anglican clergymen can.

I. The Conclusion is *No (Bishops who are not yet sufficiently senior to have a seat in the Lords) are (persons who can stand for election to the Commons).*

II. The Major Premiss is *No (Anglican clergymen) are (persons who can stand for election to the Commons).*

III. The Minor Premiss, which is of course understood though not explicit in the original statement, is *All (Bishops who are not yet sufficiently senior to have a seat in the Lords) are (Anglican clergymen).*

IV. *No Anglican clergymen are persons who . . . Commons*

───────────

[1] This examination is, of course, merely an examination of the formal validity of the argument. Formal logic is not concerned with examining the contention which the argument seeks to prove.

All Bishops who . . . Lords are members of the Anglican priesthood

∴ *No Bishops who . . . Lords are persons who . . . Commons*

V. (*a*) Structure

$\bar{M} \, e \, \bar{P}$

$\bar{S} \, a \, \bar{M}$

——   *E A E* in the First Figure.

∴ $\bar{S} \, e \, \bar{P}$

(*b*) Validity

The distribution of the Terms is as indicated in the symbolic representation, which shows that the Middle Term has been distributed at least once; and that *S* and *P*, which are both distributed in the Conclusion, are distributed in their Premisses. One Premiss is affirmative. The syllogism is valid.

Everybody with a ticket can get in, but the people in the queue can't get in, for they haven't got tickets.

  I. The Conclusion is *No* (*persons in the queue*) *are* (*persons who can enter*).

 II. The Major Premiss is *All* (*ticket-holders*) *are* (*persons who can enter*).

III. The Minor Premiss is *No* (*persons in the queue*) *are* (*ticket-holders*).

IV. *All ticket-holders are persons who can enter*
No persons in the queue are ticket-holders

  ∴ *No persons in the queue are persons who can enter*

V. (*a*) Structure

$\bar{M} \, a \, \breve{P}$

$\bar{S} \, e \, \bar{M}$

——   *A E E* in the First Figure.

∴ $\bar{S} \, e \, \bar{P}$

5

(*b*) Validity

M is adequately distributed; but *P* which is distributed in the Conclusion is not distributed in its Premiss, and the syllogism is in consequence invalid by its Fallacy of Illicit Major.

Only people with tickets can get in, so the people in the queue can't get in, for they haven't got tickets.

I. The Conclusion is *No (persons in the queue) are (persons who can enter)*.

II. The Major Premiss is *No (persons other than ticket-holders) are (persons who can enter)*.

III. The Minor Premiss is *All (persons in the queue) are (persons other than ticket-holders)*.

IV. *No persons other than ticket-holders are persons who can enter*
All persons in the queue are persons other than ticket-holders

∴ *No persons in the queue are persons who can enter*

V. (*a*) Structure

$\bar{M}\ e\ \bar{P}$
$\bar{S}\ a\ \bar{M}$ *E A E* in the First Figure.

∴ $\bar{S}\ e\ \bar{P}$

(*b*) Validity

M is distributed once at least; and both *S* and *P* which are distributed in the Conclusion are distributed in their Premisses. One Premiss is affirmative. The syllogism is valid.

This man is the murderer, for he was near the scene of the crime just about the time when it must have been committed, and an Army Service revolver that had been recently fired was found in his house (the deceased was killed by a revolver bullet of that calibre), and he had a grudge against the deceased.

I. The Conclusion is *(This man) is (the murderer)*.

II. The Major Premiss is *(The murderer) is (a man who was near the scene of the crime at the time; who has or had in his possession an Army Service revolver that had been recently fired; and who had a motive to kill the deceased).*

III. And the Minor Premiss is (*This man*) *is* (*a man who was near . . . kill the deceased*).

IV. *The murderer is a man who . . . deceased*
This man is a man who . . . deceased

∴ *This man is the murderer*

V. (*a*) Structure

$\bar{P} a \bar{M}$
$\bar{S} a \breve{M}$
——————
∴ $\bar{S} a \breve{P}$

Mood *A A A* in the Second Figure.

(*b*) Validity

S is distributed in the Conclusion, but it is distributed in its Premiss, and there is nothing wrong with the argument on that point; but *M* is undistributed in both instances, and the syllogism is consequently invalid by its Fallacy of Undistributed Middle.[1]

Army cooks must have no sense of smell, because their cooking is abominable and people without a sense of smell never make good cooks.

I. The Conclusion is *All* (*Army cooks*) *are* (*persons who have no sense of smell*).

II. The Major Premiss is *No* (*persons who have no sense of smell*) *are* (*good cooks*).

III. The Minor Premiss is *No* (*Army cooks*) *are* (*good cooks*).

---

[1] This is the syllogistic form of arguments that depend on circumstantial evidence. It is interesting to note that the argument from circumstantial evidence is always formally invalid, though it is accepted as adequate even by a court of law, provided that the description embodied in the middle term is so detailed and comprehensive that (using the same example) there is justification for asserting that it can apply to one person only, namely to the one person who is both *this man* and *the murderer*. Many scientific identifications are of this type, and depend for their persuasiveness on the quantitative exactitude of the description embodied in the middle term of what is, formally, an invalid syllogism.

IV. *No persons who have no sense of smell are good cooks*
     No Army cooks are good cooks

∴ *All Army cooks are persons who have no sense of smell*

V. (*a*) Structure

$\bar{P} \, e \, \bar{M}$
$\bar{S} \, e \, \bar{M}$

∴ $\bar{S} \, a \, \bar{P}$ *E E A* in the Second Figure.

(*b*) The Rules concerning distribution are obeyed, but both
Premisses are negative and the syllogism is therefore
invalid.[1]

---

*The existence of sensations consists in being perceived : material
objects are not sensations, therefore their existence does not
consist in being perceived.*

I. The Conclusion is *No* (*material objects*) *are* (*entities whose
existence consists in being perceived*).

II. The Major Premiss is *All* (*sensations*) *are* (*entities whose
existence consists in being perceived*).

III. The Minor Premiss is *No* (*material objects*) *are* (*sensations*).

IV. *All sensations are entities whose existence consists in being
perceived*
No material objects are sensations

∴ *No material objects are entities whose existence consists in
being perceived*

V. (*a*) Structure

$\bar{M} \, a \, \bar{P}$
$\bar{S} \, e \, \bar{M}$

$\bar{S} \, e \, \bar{P}$ *A E E* in the First Figure.

(*b*) Validity
The Middle Term is distributed once ; but *P* which is
distributed in the Conclusion is undistributed in its

---

[1] Of course, a negative premiss may be made affirmative by obversion,
and in some cases (though not in this example) a valid conclusion may
then be drawn.

Premiss. The syllogism is therefore invalid by its Illicit Process of the Major.[1]

After working through these examples the student should find little difficulty in analysing and examining any arguments that are syllogistic, or are capable of restatement in syllogistic form. This chapter on syllogism can fittingly close with a brief account of a topic that was regarded as important by most of the early logicians and by a few in later times, namely the **reduction** of syllogisms in the second, third, and fourth figures to the first or perfect figure.

Arguments in the first figure have no doubt a clarity superior to that of arguments in the other figures. A moment's thought may be necessary to see how the conclusion follows from the premisses in the latter figures, but a first figure argument is normally intelligible at sight. For reasons that are this reason otherwise expressed, Aristotle developed a system of **reducing** syllogisms in the other figures to syllogisms in the first, that is to say a system of restating them as first figure syllogisms.[2]

As the figures are distinguished by the position of the middle term in the premisses, the necessary reduction can be brought about by altering the position of the middle term. This is done by converting one of the premisses, or by transposing the premisses, or by both, to form a first figure syllogism that gives either the required conclusion, or else a conclusion from which the required conclusion can be obtained by converting. Thus the syllogism $E\,I\,O$ in the second figure, viz.

$$P\,e\,M$$
$$S\,i\,M$$
$$\therefore\ S\,o\,P$$

---

[1] The mistake here arises from failing to notice that there may be other things besides sensations whose existence consists in being perceived. Whether there actually are such is, of course, another question; but the form of the major premiss does not exclude the possibility. This kind of fallacy does sometimes occur in serious discussion.

[2] Some of his followers advanced more complicated reasons for reduction.

can be restated in the first figure by converting $P\,e\,M$ to $M\,e\,P$, which gives

$$M\,e\,P$$
$$S\,i\,M$$
$$\therefore\ S\,o\,P$$

which is the same argument in the first figure.

Similarly, $A\,E\,E$ in the second figure,

$$P\,a\,M$$
$$S\,e\,M$$
$$\therefore\ S\,e\,P$$

can be reduced by converting $S\,e\,M$ to $M\,e\,S$ and transposing the premisses, which gives

$$M\,e\,S$$
$$P\,a\,M$$
$$\therefore\ P\,e\,S$$

and this conclusion can be converted to give $S\,e\,P$.

There are some moods in which this method of direct reduction cannot be carried out, namely moods whose premisses are an A proposition and an O proposition. O cannot be converted at all, and if we convert A we get I, which would give two particular premisses, from which no valid conclusion can be drawn. Such moods can be **reduced indirectly** by using a first figure syllogism to show that the falsity of the conclusion is inconsistent with the truth of its premisses.

The somewhat narrow preoccupations of the medieval logicians led to excessive emphasis on reduction, and even to the invention of a most ingenious mnemonic that enabled the process to be carried out by rule of thumb alone.[1]   There is, however, no need

---

[1] This was devised by William Shyreswood, an Oxford man who was Chancellor of the Diocese of Lincoln in the earlier half of the thirteenth century. It passed into the logical tradition through its appearance in the *Summulae Logicales* of his later contemporary, Petrus Hispanus, who became the Pope known variously as John XX or XXI. In it each of the valid Moods was given a fabricated name, the vowels of which indicate

to memorize the methods of reduction, provided the motive that led to it and the principles on which it is based are understood. A similar comment may be made on the other masses of elaborate detail that have gathered round the doctrine of the syllogism in the twenty-three centuries during which it has had a continuous history. These elaborations have only an anti-quarian, or at the most an historic interest, and may safely for the present purpose be disregarded, provided the essentials are understood and appreciated.

NOTE ON THE DICTUM DE OMNI.

It is said by some logicians that syllogistic reasoning depends on, or proceeds in accordance with, the *Dictum de omni et nullo*, which fully expressed is :

Quod de aliquo omni dicitur (negatur), dicitur (negatur) etiam de qualibet eius parte :

What is asserted (denied) about any whole is asserted (denied) about any part of that whole.

This is debated, the answer depending on the view one takes of the nature of syllogism. That is a philosophical problem suited to a more advanced stage of study, but the student here ought to know the Dictum and to see that, at the least, it expresses a principle with which valid syllogisms are in accord.

---

the quality and quantity of its constituent propositions, while the con-sonants indicate the Figure and the treatment for Reduction if that be necessary. E.g. *Barbara (bArbArA)* is *AAA* in the First Figure ; *Bocardo (bOcArdO)* is *OAO* in the Third Figure, and the medial *c* indicates that it can be reduced only by the process of indirect reduction. The com-monest version runs as follows :

> *Barbara Celarent Darii Ferioque prioris ;*
> *Cesare Camestres Festino Baroco secundae ;*
> *Tertia Darapti Disamis Datisi' Felapton*
> *Bocardo Ferison habet ; quarta insuper addit*
> *Bramantip Camenes Dimaris Fesapo Fresison.*

As an illustration of the general familiarity with the traditional logic, it is worth noting that in Oxford the name Bocardo was quaintly used of the gatehouse and prison in which, for nearly five hundred years after Shyreswood's time, successive Vice-Chancellors incarcerated disorderly members of their university. Not all of these, the reader may be interested to know, were undergraduates.

CHAPTER V

HYPOTHETICAL ARGUMENT

THE arguments that we have so far considered are categorical, i.e. they consist of statements which, whether true or false, are categorical or unconditional. There is, however, another and very common type of argument that is partly conditional or hypothetical in character, such as the following:

If the tank is empty, the car will not start. It is empty, therefore the car will not start.

If this bookcase is less than seven feet high, it will pass through the study door. It is less than seven feet high, therefore it will pass through.

These **hypothetical arguments,** as they are called, can be regarded as consisting of propositions standing in a certain kind of systematic relation to each other, just as syllogisms can be regarded as consisting of propositions standing in another kind of systematic relation to each other.

The nature of that systematic relation in which the propositions constituting an hypothetical argument stand to each other will become clear if the examples quoted are rewritten in the following way:

If *the tank is empty* [1]   **then** *the car will not start*
The tank is empty

∴ *The car will not start*

[1] For the sake of brevity and convenience these propositions are left in their colloquial form (instead of being restated in logical form) as we are here concerned with their relation to each other, and not with their internal structure.

If *this bookcase is less than*   **then** *it will pass through the*
   seven feet high   *study door*

This bookcase is less than seven feet high

∴ *It will pass through the study door*

When discussing syllogism, we advanced by a series of steps to the discovery that there is a form of syllogism that can be examined by itself, without reference to any specific meaning that the propositions involved may convey.[1] By similar steps we can advance to the discovery that there is a form of hypothetical argument that can be examined by itself without reference to any specific meaning that the propositions involved may convey.

By using the familiar symbolism, a hypothetical argument can be represented as follows :

$$\text{If } A\,a\,B, \quad \text{then} \quad C\,a\,D$$
$$A\,a\,B$$

$$\therefore C\,a\,D$$

A further simplification can be made by using a single symbol to represent a complete proposition, instead of representing only a term or the copula thereof.   If p represents the proposition $A\,a\,B$, and q represents the proposition $C\,a\,D$, then a typical hypothetical argument can be symbolized thus :

$$\text{If } p \quad \text{is true, } \textbf{then} \quad q \quad \text{is true,}$$
$$p \quad \text{is true,}$$

$$\therefore q \quad \text{is true}$$

or, even more simply,

$$\text{If } p, \quad \text{then} \quad \textbf{q}$$
$$p$$

$$\therefore \textbf{q}$$

[1] See page 45, last line.

In an argument of this form the conclusion follows from the premisses, no matter what specific meaning be given to the propositions represented by p and q. I.e. what we are dealing with here is the *form* of hypothetical argument.

Just as there are different kinds or modes or moods of syllogism, so there are different kinds or modes or moods of hypothetical argument. For instance, the two examples already quoted are in one mood, and the following is in another:

If this apple is ripe, it will be easy to pluck. This apple is not easy to pluck, therefore it can't be ripe.

If *this apple is ripe*   **then** *it is easy to pluck*
              It is not easy to pluck

∴ *It is not ripe*

If p *is true,* **then** q *is true*   *If* p, *then* q
      q *is not true*   *not-q* [1]

∴ p *is not true*   ∴ *not-p*

These different kinds of examples show how propositions are related to each other to form hypothetical arguments. Further light is thrown on the structure of hypothetical arguments by regarding them, from a slightly different viewpoint, as consisting like syllogisms of two premisses and a conclusion, though of course the premisses differ both in composition and interrelation from the premisses of syllogism.

The Major Premiss [2] takes the form *If p, then q*, i.e. it asserts a conditional or hypothetical relation between two propositions, such that if the *antecedent proposition* is true, then the *consequent proposition* is true also.

[1] *not-q* is used to symbolize the contradictory of q, i.e. to symbolize the denial of q. This kind of symbol must not be confused with the *not-S* kind (p. 39), in which S represents a term and not a proposition.

[2] This corresponds to the use of the names major and minor in the doctrine of the syllogism, for in most cases the major premiss states the general principle or rule of which the minor premiss states an instance.

the Minor Premiss   is *one* of these propositions (either in its affirmative or its negative form) (e.g. p)

and the Conclusion   is *the other* of these propositions (either in its affirmative or its negative form) (e.g. $\therefore q$).

Given an argument beginning *If p, then q*, the next step must, naturally, be either to affirm the antecedent [1] (e.g. *The tank is empty*) or to deny it; or to affirm the consequent, or to deny it (e.g. *It is not easy to pluck*). In other words, the minor premiss must be either the affirmation of the antecedent, or the denial of the antecedent, or the affirmation of the consequent, or the denial of the consequent. These four possibilities give rise to the four kinds or modes or moods of the hypothetical argument. We shall now examine these four moods in turn, with special reference to their validity.

AFFIRMING THE ANTECEDENT

$$If\ p,\ then\ q$$
$$p$$
$$\therefore q$$

This mood is exemplified by the arguments already quoted on page 64, or by the following:

If the tide is ebbing, it is dangerous to attempt to pass the narrows. The tide is now on the ebb, so it is dangerous to make the attempt.

If *the tide is ebbing*   **then** *it is dangerous to attempt to pass the narrows*

       The tide is ebbing

    $\therefore$ *It is dangerous to attempt to pass the narrows*

[1] **A** shorter name for the antecedent proposition

The validity of this very common mood, namely the **affirming of the antecedent,** requires no further discussion.

DENYING THE ANTECEDENT

If p, then q
not-p

---

which does not justify any conclusion. If it is not at once evident that the conclusion *not-q* which is sometimes alleged from the above type of argument is invalid, an example will make it so.

If I am in Edinburgh, then I am in Scotland. I am not at present in Edinburgh, therefore I am not at present in Scotland

If *I am in Edinburgh* **then** *I am in Scotland*
I am not in Edinburgh

∴ *I am not in Scotland*

This is obviously invalid, for I might be in Glenlyon or the Mull of Galloway or in innumerable other places that are not Edinburgh, and yet be in Scotland. There are many possibilities (e.g. that I am in Glenlyon; or in Galloway; or anywhere else in Scotland outside Edinburgh), all of which are compatible with the denial of the antecedent (i.e. compatible with *I am not in Edinburgh*). The argument gives no information as to which of those alternative possibilities is the actual state of affairs, and thus there is no ground for asserting any one of those possibilities as a conclusion. In other words, the denial of the antecedent is a statement that is compatible with a variety of alternative possibilities. It gives no ground for differentiating among them, and therefore no conclusion can be drawn. The argument by denying the antecedent is invalid.

DENYING THE CONSEQUENT

$$If \; p, \; then \; q$$
$$not\text{-}q$$

$$\therefore \; not\text{-}p$$

As an example of this very common mood consider the kind of argument that disproves an hypothesis by showing that some of the consequences of that hypothesis are contrary to fact.

If your theory of the foreign exchanges is sound, then the Bank Rate ought to have gone down on the 1st of June, but it did not, so your theory must be wrong.

If *your theory of the foreign*   **then** *the Bank Rate went down*
    exchanges is sound   *on the 1st of June*

The Bank Rate did not go
down on the 1st of June

∴ *Your theory of the foreign exchanges is not sound*

Argument in this mood, namely by **denying the consequent,** is clearly valid.

AFFIRMING THE CONSEQUENT

$$If \; p, \; then \; q$$
$$q$$

which does not justify any conclusion. If it is not at once evident that the conclusion p, which is sometimes alleged from the above type of argument, is invalid, an example will make it so.

If Smith did not sit the exam., then he did not pass. He has not passed, therefore he did not sit.

If *Smith did not sit the exam.* **then** *Smith has not passed the exam.*

Smith has not passed the exam.

∴ *Smith did not sit the exam.*

This is obviously an unreasonable conclusion, for the bare statement that he did not pass the examination does not tell us whether he sat it or not. There are many possibilities (e.g. that he sat; that he did not sit; that he left after half an hour; and so forth), all of which are compatible with the affirmation of the consequent (i.e. are compatible with *He has not passed*). The argument gives no information as to which of these alternative possibilities is the actual state of affairs, and thus there is no ground for asserting any one of these possibilities as a conclusion. In other words, the affirmation of the consequent is a statement that is compatible with a variety of alternative possibilities. It gives no ground for differentiating among them, and therefore no conclusion can be drawn. The argument by affirming the consequent is invalid.

To sum up, **affirming the antecedent** and **denying the consequent** afford valid conclusions, and the other two moods do not.

At this point some examples can profitably be dealt with, that is, they can be expressed explicitly as hypothetical arguments, their structure and validity being examined. As in the similar treatment of syllogistic arguments, tacit omissions may have to be supplied, and considerable restatement may be necessary, in order to make clear the constituent propositions and the relations in which they stand to each other.

Care is necessary to distinguish between the *truth of the conclusion* and the *validity of the argument*, which are often confused.[1]

[1] See note 2 on page 33.

EXAMPLES OF THE ABOVE

If this is the right key it will open the lock. It is not the right key, therefore it will not open the lock.

If *this is the right key*   **then** *it will open the lock*
          It is not the right key

          ∴ *It will not open the lock*

If p, then q
     not-p

∴ *not-q*   Denial of the Antecedent, and thus invalid.

The lock might be a cheap one, and a key intended for another lock might fit it also.

The colloquial expressions of hypothetical arguments often employ conjunctions other than *if*, such as *when, whenever, wherever, unless*. These have to be replaced by an *if* construction, special care being taken with conjunctions like *unless* that involve a negative significance.

The wireless set will not work; for it won't work unless it has been repaired, and it hasn't.

The *unless* clause should be restated in the conventional *if* form, thus :

The wireless set will not work, for if it has not been repaired it will not work, and it has not been repaired.

If *the wireless set has not*   **then** *it will not work*
          been repaired

          The wireless set has not been repaired

          ∴ *It will not work*

It is to be noticed that a proposition, though containing a negative, may be an affirmation and not a denial, if the proposition of which it is the affirmation itself contains a negative. In the above example the minor premiss is a negative proposition, but it is the affirmation and not the denial of the antecedent. The symbolic representation helps to make this clear.

If p, then q
   p
  ──
∴ *q*   Affirmation of the Antecedent. Valid.

When a patient has malaria, his temperature ' swings ' (i.e. it rises and falls suddenly). This patient has a swinging temperature so he must have malaria.

If *the patient has malaria*   **then** *he has a swinging temperature*

He has a swinging temperature

∴ *The patient has malaria*

If p, then q
   q
  ──
∴ *p*   Affirmation of the Consequent and thus invalid.

There might be many causes other than malaria for a swinging temperature. Though the argument is worthless as a *proof*, yet it is a *verification*, and an accumulation of similar arguments would no doubt convince the doctor that the patient had malaria (cf. page 59 n.), provided that there was not a single contrary indication (cf. page 69).

This writer cannot be a materialist, for a materialist has to be a determinist, and he is not a determinist.

The first step is to restate this in the conventional *if* form, thus:

This writer cannot be a materialist, for if he is a materialist he is a determinist, and he is not a determinist.

If *this writer is a materialist*   **then** *he is a determinist*

He is not a determinist

∴ *This writer is not a materialist*

If p, then q
  not-q

∴ *not-p*   Denial of the Consequent. Valid.

CHAPTER VI

DISJUNCTIVE ARGUMENT. MORE COMPLEX ARGUMENTS

ALL the principal familiar types of argument have now been discussed except one, namely arguments of the " either . . . or . . ." type, such as the following :

This liquid is either an acid or an alkali. It is not an acid, so it must be an alkali.

The man in the scarlet gown is a D.Litt., for he is either a D.Litt. or a D.Sc., and he is not a D.Sc.

These **disjunctive arguments,** as they are called, can be regarded as consisting of propositions standing in a certain kind of systematic relation to each other, exactly as syllogisms and hypothetical arguments can be regarded as consisting of propositions standing in other kinds of systematic relation to each other.

The nature of the systematic relations in which the propositions constituting a disjunctive argument stand to each other will become clear if the examples quoted are rewritten in the following way :

Either *this liquid is an acid*   **or** *this liquid is an alkali*

This liquid is not an acid

∴ *This liquid is an alkali*

Either *the man in the scarlet*   **or** *the man in the scarlet gown gown is a D.Litt.*   *is a D.Sc.*

The man in the scarlet gown is not a D.Sc.

∴ *The man in the scarlet gown is a D.Litt.*

Just as syllogisms and hypothetical arguments have a form that can be examined without reference to any specific meaning that the propositions involved may convey, so also disjunctive arguments have a form that can be examined without reference to any specific meaning that the propositions involved may convey. If we again use a single symbol for each of the propositions involved, then a disjunctive argument can be represented thus :

Either *p* *or* *q*
     not-p   **or**   *not-q*
     ——   ——
   ∴ *q*   ∴ *p*   (The order of the alterna-
                         tives makes no difference.)

In an argument of this form, the conclusion follows from the premisses, no matter what specific meaning be given to the propositions represented by *p* and *q*. *I.e.* what we are dealing with here is the *form* of disjunctive argument.

These examples show how propositions are related to each other to form disjunctive arguments. Further light is thrown on the structure of disjunctive arguments by regarding them, from a slightly different viewpoint, as consisting, like syllogisms and hypothetical arguments, of two premisses and a conclusion, though of course the premisses differ both in composition and inter-relation from the premisses of syllogism and the premisses of hypothetical argument.

The Major Premiss takes the form *Either* *p* *or* *q*, i.e. it asserts a disjunctive relation between two alternative propositions.

The Minor Premiss is one of these propositions (either in affirmative or its negative form) (e.g. *not-p*).

The Conclusion is the other of these propositions (either in its affirmative or its negative form) (e.g. ∴ *q*).

A difficulty arises here because the ordinary English usage of

" either . . . or " is ambiguous. When we say " X is either Y or Z ", we may mean

    (a) that it is either the one or the other, but cannot be both (*This liquid is either an acid or an alkali*). This is commonly called an " exclusive " or **strong** disjunction;

or (b) that it is either the one or the other and, for all we know, may be both (*This man is either a D.Litt. or a D.Sc.*). This is commonly called a **weak** disjunction.

In ordinary speech we mostly use the same form of words for both cases, relying on our hearer's common sense to ensure that he does not misunderstand. If a child is told by his father that he can have either a train or a box of lead soldiers for Christmas, he understands without further specification that he cannot have both, *i.e.* that his father intends a strong disjunction. If, on the other hand, his father looks out of the window and says " It may rain this afternoon, or it may be windy ", then he understands without further specification that his father intends a weak disjunction, *i.e.* that it may be both wet and windy.

In logic, on the other hand, every disjunction should be stated in a way which indicates explicitly whether it is intended to be strong or weak. The tradition is not uniform on this point, but the majority of logicians would approve of the practice of using " *either . . . or* " with the addition of " *but not both* " for a strong disjunction, and " *either . . . or* " for a weak disjunction, *i.e.* for the minimum meaning these words have in ordinary speech, leaving it open whether the alternatives are or are not both true.

Given a disjunction (which may be strong or weak), then the next step in the argument must be either to affirm one of these alternatives,[1] or to deny one of these alternatives ; *i.e.* the minor premiss must, naturally, be either the affirmation or the denial of one of the alternatives. These two possibilities give rise to the two kinds of modes or moods of the disjunctive argument, and each of these may be found with a weak or a strong disjunction, thus giving rise to four possible cases. We shall now examine these four cases in turn with special reference to their validity.

[1] A shorter name for the *alternative proposition*.

DENYING AN ALTERNATIVE (Weak disjunction)

Either p or q
not-p

∴ *q*

*This man is either a Justice of the Peace or a Privy Councillor,
but he is not a Privy Councillor so he must be a J.P.*

Either *this man is a J.P.* **or** *this man is a Privy
Councillor*

He is not a Privy Councillor

∴ *He is a J.P.*

The validity of this very common mood, namely argument
by denying an alternative in the case of a weak disjunction,
needs no further discussion.

DENYING AN ALTERNATIVE (Strong disjunction)

Either p or q (but not both)
not-p

∴ *q*

*This is either a First Class compartment or a Third Class one.
It is not a First, so it must be a Third.*

Either *this is a First Class* **or** *this is a Third Class
compartment* *compartment*

This is not a First Class compartment

∴ *This is a Third Class compartment.*

The validity of this very common mood, namely argument
by denying an alternative in the case of a strong disjunction,
needs no further discussion.

AFFIRMING AN ALTERNATIVE (Weak disjunction)

$$\textit{Either p or q}$$
$$\underline{p}$$

which does not justify any conclusion. If it is not at once evident that the conclusion *not-q*, which is sometimes alleged from the above type of argument, is invalid, an example will make it so.

This substance contains either carbon or oxygen. It contains carbon, therefore it does not contain oxygen.

Either *this substance con-*   **or** *this substance contains*
tains carbon   *oxygen*

This substance contains carbon

∴ *This substance does not contain oxygen*

This alleged inference is plainly absurd, for the disjunction is weak, i.e. the substance might very well contain both. There are various possibilities (e.g. that the substance contains oxygen ; that it does not contain oxygen), all of which are compatible with the affirmation of the alternative (i.e. are compatible with *This substance contains carbon*) and the argument gives no information as to which of these possibilities is the actual state of affairs, and thus there is no ground for asserting any one of these possibilities as a conclusion. In other words, the affirmation of an alternative when the disjunction is weak is a statement that is compatible with a variety of possibilities. It gives no ground for differentiating among them, and therefore no conclusion can be drawn.

Affirming an alternative in the case of a weak disjunction is invalid.

AFFIRMING AN ALTERNATIVE (Strong disjunction)

$$Either\ p\ or\ q\ (but\ not\ both)$$

$$\frac{p}{\therefore\ not\text{-}q}$$

This is either a turnip or a carrot and it is a turnip, so it is not a carrot.

Either *this is a turnip*   or *it is a carrot (but not both)*

$$\frac{It\ is\ a\ turnip}{\therefore\ It\ is\ not\ a\ carrot}$$

The disjunction is strong, i.e. the alternatives cannot both be true. If one of them is true, the other must therefore be false. Affirming an alternative, in the case of a strong disjunction, is valid.

It must be borne clearly in mind that our knowledge that a disjunction is weak or strong (if we have it) comes from special knowledge of the subject matter under discussion. From the argument alone we cannot tell whether the disjunction is weak or strong. In purely formal arguments, such as the simple ones

(*a*) *Either p or q*   (*b*) *Either p or q*
  not-p   *p*
$$\therefore\ q$$ $$\therefore\ not\text{-}q$$

we have to interpret the disjunction as weak, because there is nothing in the form of the argument alone to tell us whether the alternatives are or are not exclusive. In such purely formal arguments as (*a*) and (*b*) above, (*a*) is therefore valid and (*b*) invalid.

To sum up, a valid conclusion to a disjunctive argument can always be obtained by denying an alternative. In the case of a strong disjunction, a valid conclusion can in addition be obtained by affirming an alternative.

In restating colloquially expressed arguments as explicit

disjunctive arguments the same considerations apply as in the similar treatment of syllogisms and hypothetical arguments.

EXAMPLES OF THE ABOVE

Either the cash in hand tallies with the accounts, or else the accounts are wrong. But it does, so the accounts are correct.

Either *the cash in hand tallies*   **or** *the accounts are not correct*
        with the accounts

        The cash in hand does tally with the accounts

   ∴ *The accounts are correct*

Either p or q (or both)
        p

   ∴ *not-q*   Affirming an Alternative with a weak disjunction. Invalid.

The cash might tally with the accounts, but the accounts are not thereby proved correct, for if equal omissions had been made from both Income and Expenditure, the cash would still tally, yet the accounts would be wrong.

This is either a briar pipe or a meerschaum. It is a briar, so it is not a meerschaum.

Either *this is a briar pipe*   **or** *this is a meerschaum pipe*

        This is a briar pipe

   ∴ *This is not a meerschaum pipe*

Either p or q (but not both)
        p

   ∴ *not-q*   Affirming an Alternative with a strong disjunction. Valid.

He's a rogue or I'm a Dutchman.

This is, of course, an ellipsis for *Either he is a rogue or I am a Dutchman, and I am not a Dutchman, therefore he is a rogue.*

Either *he is a rogue*   **or** *I am a Dutchman*

           I am not a Dutchman

         ∴ *He is a rogue*

Either p or q

   not-q

  ∴ p Denying an Alternative. Valid.

As a further example of the need to discriminate between truth and validity, consider the following :

The verdict must be either ' Guilty ' or ' Not Guilty '. It isn't Guilty ', so it must be ' Not Guilty '.

Either *the verdict is ' Guilty '*   **or** *the verdict is ' Not Guilty '*

          The verdict is not ' Guilty '

        ∴ *The verdict is ' Not Guilty '*

Either p or q

   not-p

  ∴ q Denying an Alternative. Valid.

It might be pointed out that in Scots Law this is not the case. If a verdict is not ' Guilty ', it does not follow that it is ' Not Guilty ', for it might be ' Not Proven '. This does not mean that the argument is valid if we think of English Law, and invalid if we think of Scots Law. The argument is a formally valid argument, and cannot be anything else. What is questionable is not the validity of the argument, but the truth or falsity of the major premiss, for it is true in reference to English Law but false in reference to Scots Law, since in the latter it is not the case that a verdict is either ' Guilty ' or ' Not Guilty ', for it might be ' Not Proven '.

Disjunctive argument is the last of the four types of inference recognized by the tradition, the others being, of course, Immediate Inference, Syllogism, and Hypothetical Argument.

Chains of reasoning which are apparently different from or more complex than any of these four can be treated as consisting of combinations of two or more inferences, each of which is an example of one or other of the four types. For instance, the following brief but highly complex chain of reasoning consists of two syllogisms, one of them providing a conclusion that then serves as a premiss in the other.

> *Brutus says that Caesar was ambitious ;*
> *And Brutus is an honourable man.*

The first syllogism shows that what Brutus says is true, and this conclusion is then used as a premiss in a second syllogism to prove that Caesar was ambitious. It is not relevant that the whole argument is employed ironically in the speech in which it occurs.

I.   The first stage of the argument is intended to show that what Brutus says is true, i.e. the Conclusion is *All* (*statements made by Brutus*) *are* (*true statements*). Note that the Subject is not *Brutus* but *what Brutus says*, i.e. the logical subject is not the same as the grammatical subject of the sentence.

II.   The Major Premiss, which is left to be understood, is *All* (*statements made by an honourable man*) *are* (*true statements*).

III. The Minor Premiss is *All* (*statements made by Brutus*) *are* (*statements made by an honourable man*). Here again the Subject and Predicate require drastic reformulation.

IV.   (*a*) Structure
      All statements made by an honourable man are true statements
      All statements made by Brutus are statements made by an honourable man

$\therefore$ *All statements made by Brutus are true statements*

$$\bar{M} \, a \, \breve{P}$$
$$\bar{S} \, a \, \bar{M}$$
$$\rule{2cm}{0.4pt}$$
$$\therefore \, S \, a \, \breve{P}$$ *A A A* in the First Figure.

(b) Validity

This is clearly valid.

The second part of the argument utilizes that Conclusion as its Major Premiss:

I. The Conclusion of the second part, and of the whole argument, is (*That Caesar was ambitious*) *is* (*a true statement*).

II. The Major Premiss is the Conclusion of the first syllogism, *All* (*statements made by Brutus*) *are* (*true statements*).

III. The Minor Premiss is (*That Caesar was ambitious*) *is* (*a statement made by Brutus*).

IV. (a) Structure

All statements made by Brutus are true statements
That Caesar was ambitious is a statement made by Brutus

∴ *That Caesar was ambitious is a true statement*

$$\bar{M} \, a \, \breve{P}$$
$$\bar{S} \, a \, \bar{M}$$
$$\overline{\phantom{\bar{S} \, a \, \breve{P}}}$$ *A A A* in the First Figure.
$$\therefore \bar{S} \, a \, \breve{P}$$

(b) Validity

This also is clearly valid.

As a further instance of these simple principles exemplified in a highly complex argument, consider the following: *If Government Departments are allowed to issue regulations having the force of law, the private citizen will have no redress against interference. If the Departments are not allowed to do so, then Parliament will become congested by more business than it can possibly cope with. But the Departments must be either so allowed or not allowed, and in consequence either the private citizen will have no redress against bureaucratic interference, or Parliament will become hopelessly congested.*

If p, then q, and if not-p then r
But either p or not-p
∴ Either q or r.

This is an example of a **dilemma**, a form commonly used

in polemic to force an opponent to admit some distasteful conclusion. It is an argument consisting of a disjunction, and of
hypotheticals that connect both the alternatives of the disjunction with a conclusion or conclusions.

The early logicians gave names to some of the commoner
complex inferences, such as the dilemma, but there is no special
need to learn and remember them. What is important to
learn and remember is the doctrine that the most complicated
reasoning consists of comparatively simple inferences, and that
each of these inferences is either an immediate inference, or
a syllogism, or a hypothetical argument or a disjunctive
argument.[1]

We are now in a position to appreciate adequately the remarkable simplification on which the traditional doctrine insists.

It maintains the convention

(a) that all reasoning, whether long or short, simple or
complex, can be analysed and treated as consisting of
an inference, or combination of inferences, each of which
must be one or other of only four types, viz. :

immediate inference
syllogism
hypothetical argument
disjunctive argument ;

(b) that each of these inferences can be analysed and treated
as consisting of propositions which stand to each other
in one or other of a strictly limited number of systematic
relations, viz. :

The moods of immediate inference [2]
The moods of syllogism
The moods of hypothetical argument
The moods of disjunctive argument ;

---

[1] It is, of course, equally important to remember that this is only a
convention carried over from the ancient and medieval views on logic.
Few philosophers to-day would accept it.

[2] Conversion, obversion, &c.

(c) that each of these propositions can be analysed and treated as consisting of a subject term and a predicate term standing in one or other of only four relations, viz. :

> universal and affirmative
> universal and negative
> particular and affirmative
> particular and negative.

And as a useful corollary, it has been possible to classify fallacies, i.e. arguments that may appear valid but are not, and to give them technical names such as illicit process or undistributed middle.[1]

ADDENDUM

This deals with developments beyond what are generally regarded as the limits of the traditional teaching, but it is both useful and suggestive at this stage to see how they follow from it. The reader cannot have failed to notice in passing that the inferences on pages 71–73, which are there expressed as hypothetical arguments, could have been expressed as syllogisms, thus suggesting that the same reasoning can be expressed either in syllogistic or in hypothetical form. For instance, the following argument,

This writer cannot be a materialist, for a materialist has to be a determinist, and he is not a determinist,

can be expressed as a syllogism thus :

I. The Conclusion is *This writer is not a materialist.*

II. The Major Premiss is *All materialists are determinists,* which is the Logical Form of the clause *a materialist has to be a determinist.*

III. The Minor Premiss is *This writer is not a determinist.*

[1] These strictly formal fallacies are the only kinds of mistake in reasoning that can be detected solely by a knowledge of formal logic. Some accounts of the traditional logic discuss other kinds of looseness of thought, but any adequate treatment of these would raise wider issues than the strictly formal problems dealt with here.

IV. (a) Structure

> *All materialists are determinists*
> *This writer is not a determinist*

∴ *This writer is not a materialist*

$$\bar{P} \ a \ \bar{M}$$
$$\bar{S} \ e \ \bar{M}$$
$$\overline{}$$
∴ $\bar{S} \ e \ \bar{P}$ *A E E* in the Second Figure.   Valid.

As another example, the reasoning expressed as a valid syllogism on page 58 can be expressed as a valid hypothetical argument thus :

If *these persons have not*   **then** *they can not enter*
tickets

         These persons have not tickets

     ∴ *They cannot enter*

∴ *If p, then q*
   p
   ‾‾
∴ *q*   Affirmation of the Antecedent.   Valid.

Similarly the reasoning expressed as an invalid syllogism on page 58 can be expressed as an invalid hypothetical argument thus :

If *this man is the murderer*   **then** *he was near the scene . . .*
                                       sufficient motive to kill the
                                       deceased.

He was near the scene . . . sufficient motive to kill the deceased

∴ *This man is the murderer*

If p, then q
   q
   ‾‾
∴ *p*   Affirmation of the Consequent.   Invalid.[1]

[1] Cf. page 72, last para.

It will be found that with some ingenuity the other syllogisms also can be expressed as hypothetical arguments. This illustrates that it is possible to express what is fundamentally the same reasoning either in the syllogistic or in the hypothetical form, the valid moods of the one corresponding to the valid moods of the other, and the invalid moods of the one corresponding to the invalid moods of the other.[1]

Moreover, it is further possible to express the same reasoning as a disjunctive argument also. For instance, the reasoning expressed as a syllogism on page 58 and as an hypothetical argument on page 86 can be expressed as a disjunctive argument thus :

Either *these persons have*   **or** *they cannot enter*
    tickets

        These persons have not tickets
       ───────────────────────
∴ *They cannot enter*

Either p or q
   not-p
   ─────
∴ *q*   Denial of an Alternative. Valid.

It is illuminating to discuss this with reference to the bare form of the arguments, i.e. by considering the symbolic representation alone. Given the following hypothetical argument,

$$If \ p, \ then \ q$$
$$not\text{-}q$$
─────
$$\therefore \ not\text{-}p$$

───────────

[1] Some logicians maintain that all arguments are essentially syllogistic and that an hypothetical argument is only a way of expressing what is ' really ' a syllogism. Others maintain the contrary, and others again regard both forms as differing only in manner of expression, neither having any priority as the more fundamental. This is another of those philosophical problems that the study of formal logic raises without answering.

consider what would be the disjunctive argument corresponding to it.

The major premiss *If p, then q* can be expressed as a disjunctive statement thus,

Either q or not-p

[because *If p then q* means that we cannot have *p* without as a consequence having *q*.

i.e. if we do not have *q* we cannot have *p*.

i.e. if we do not have *q* we must have *not-p*.

i.e. *Either q or not-p*.]

The minor premiss is the same as in the hypothetical form, and the conclusion is the same as in the hypothetical form.

The hypothetical argument can then be expressed as a disjunctive argument thus:

If p, then q	*Either q or not-p*
not-q	*not-q*
∴ *not-p*	∴ *not-p*

Similarly the other moods of hypothetical argument can be expressed as disjunctive arguments also.

This discussion suggests questions that will have to be considered in any inquiry into the nature of logical form. It also suggests the beginning of a development of which the student will hear more at a later stage, if he continues the study of logic, namely ' symbolic logic ', as it is now called. By considering propositions as units and representing them by single symbols, and by inventing other symbols to represent the *if–then* relation and the *either–or* relation and the like, it is possible to develop highly complex systems, in appearance not unlike algebra, of propositions in all sorts of relation to each other. This will be found worked out in any of the text-books of symbolic logic

CHAPTER VII

LOGICAL DIVISION AND DEFINITION

OF the essentials of the traditional logic there now remains to be discussed only the conventional doctrine of Logical Division and Definition. This is most easily understood as a corollary to the doctrine of the denotation and connotation of terms.

We found that in most, though possibly not in all cases, the denotation and the connotation of a term vary inversely. If the connotation be progressively increased by the addition of qualifying epithets, then the denotation is progressively decreased by corresponding stages. This may be described from a slightly different point of view by saying that the original numerous denotation of the terms has been divided and subdivided into several smaller classes, each of them being the denotation corresponding to a certain stage of qualification of the connotation. If for example the original term be *soldier*, and if we progressively qualify it by adding the epithets *regular* and *cavalry* thereby increasing the connotation in two stages, then we have at the same time divided the large class of entities which was the denotation of *soldier* into three sub-classes, corresponding to stages in the increase of the connotation.

To show this more clearly, take a diagram to represent the denotation of *soldier*, thus:

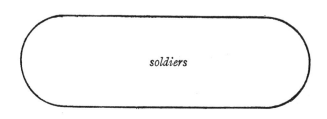

soldiers

7

By introducing the qualification *regular*, we divide the class *soldiers* thus :

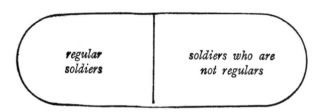

and by adding the further qualification *cavalry*, we further subdivide the class *regular soldiers* thus :

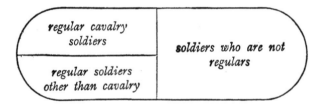

An alternative diagram would depict the same situation in this way :

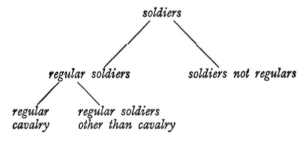

This process of classifying, or of dividing a class into its constituent sub-classes and these again into their constituent sub-classes, is known technically as **division,** and this we are now to examine.

We habitually classify and 'divide' in order to gain clarity and comparative simplicity, whether the classification be carried out in a small sphere for practical convenience, as in dividing a pile of letters into *Answered* and *Unanswered*, or whether it be carried out in systematic detail for scientific or other important purposes, as in Linnæus's classification of plants or in the lawyer's classification of kinds of ownership.

Both for theoretical and for practical needs the principles according to which this process of division can properly be carried out require to be examined and must be understood, because competent classification is essential to intellectual advance, and irresponsible classification leads to endless confusion. We shall now consider some examples that may suggest what these principles are.

If I divide *soldiers* into *officers* and *artillerymen*, then instead of classifying I have created confusion, for these two sub-classes overlap, the officers of the Royal Artillery being members of both. But it would be a clarifying simplification to divide *soldiers* into *officers* and *other ranks*, or into *artillerymen* and *members of the other arms of the Service*. In both these latter cases the two sub-classes do not overlap and no member of one class is a member of the other also ; which suggests as a principle that the **sub-classes in the division of a class must be mutually exclusive.** A little further reflection shows that this must be so, for otherwise we have a sort of ' cross division ' that is not division at all.

The other principle may likewise be suggested by an example. If I make a classification of soldiers as *cavalry* and *artillery* and leave it at that, then I have again confused the situation, this time by omitting the infantry and several other branches. To make a satisfactory division I ought to take account of them all, classifying *soldiers* as *cavalry, artillery, infantry* and *members of the remaining branches* ; which would be a classification exhausting the class *soldiers*. This suggests that the **sub-classes in the division of a class must together make up the whole class,** which is a principle that hardly needs any further thought for its justification.

A logical division or classification must then accord with

these two principles, and the penalty for disregarding either of them is confusion.

It is customary in the traditional logic to use the technical names **genus** and **species,** calling the class that is to be divided the **genus,** and calling the sub-classes thereof the **species** of that genus. These names are relative to each other in their significance, for any class can be called either a genus or a species according to the way we look at it. If we regard it as a class to be divided into sub-classes, then it is a genus and the sub-classes are its species ; but if we regard it as part of a wider class, then it is a species of that wider class, which is its genus.[1]

Of course, division and classification are always to some extent arbitrary, the classes and divisions depending partly on the subject matter and partly on the standpoint that the classifier consciously or unconsciously adopts, with the consequence that difficulties arise. For instance, there will be ' borderline cases ' to cope with whenever it is difficult to find sub-classes that are mutually exclusive. This does not mean that classifying should be abandoned as hopeless, but that some better means of distinguishing the sub-classes should be devised so that there are no borderline cases. In other words, **a formulation must be devised such that all the sub-classes are mutually exclusive.**

This process of dividing and classifying can in theory be extended in both directions, until we have a scheme with the widest possible or conceivable class at the top, divided and subdivided until at the bottom we have particular individuals. Many painstaking examples occur in the older text-books, but they are not important. It is sufficient if the student understands that any classification or division must, if it is not to

[1] The usage of the words ' genus ' and ' species ' in Zoology (and Botany) is exceptional, and may be confusing unless distinguished as a special case. Though any zoological class is, logically considered, a genus in reference to inferior classes and a species in reference to superior classes, yet zoologists have in the main agreed to apply the name genus to one kind of class only, and the name species to its immediate sub-classes only.

be merely confusing, be such that all the species are mutually exclusive and that all the species together make up the whole genus.

This mode of approaching the question of classification by regarding it as a development from the doctrine of denotation and connotation is selected to make it easier to understand, and must not be taken to imply that classifying is a secondary or derivative process dependent on our accepting the doctrine of denotation and connotation. It would be equally possible, though much less simple, to examine classification and division first of all, and then go on to discuss denotation and connotation as a corollary thereof. All these topics are integrally connected, and each of them acquires a fuller significance when treated in relation to, or even in terms of, the others.

By a further application of this method, the theory of logical division and classification may be used to lead up to and explain definition.

An example of a satisfactory division would be the division of the genus *rectilineal figures* into the species *triangles, quadrilaterals, pentagons, hexagons* and *others*.

A diagram shows the situation thus :

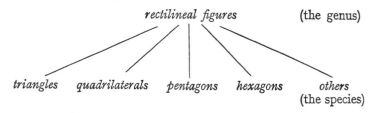

rectilineal figures (the genus)

triangles   quadrilaterals   pentagons   hexagons   others
(the species)

If I am asked to define *hexagons*, it seems natural to begin by saying that they are *rectilineal figures*, and to complete the definition by adding that they are *six-sided* rectilineal figures. That is, I begin by stating the genus to which hexagons belong, and then specify which species of that genus they are, by pointing out that they differ from all other species of that genus in having six sides. In other words, I state the genus of which the entities to be defined are a species, and then specify which species by mentioning the one quality or attribute that differentiates that species from all the rest of the genus. Such a quality serving to differentiate one species from all the rest of the same genus, is technically called a **difference,** and the process of defining that we are discussing is in consequence fittingly called **definition by genus and difference.** A glance at any page of a dictionary will show that most explanations of the meaning of words—apart from those consisting merely in the substitution of synonyms—are definitions of this type. A definition, then, is dependent upon a division, whether this division be made explicitly with a view to definition, or whether it be tacitly presupposed and taken for granted.

If a definition is to be adequate, the division it depends on must have been made in accordance with the two principles with which we are already familiar, and in addition the genus must be clearly stated and the difference must be adequate to specify exactly which of the species is intended.

The numerous ' Rules of Definition ' given in the old logic books are rules of thumb designed to ensure that the above requirements, together with some other common-sense conditions, are satisfied, and there is no need to memorize them, provided that the theory of definition here discussed is understood and applied.

The giving of a technical name (i.e. difference) to one kind of quality leads to the consideration of other kinds and to the attempt to distinguish and name them. This has been undertaken by logicians in various and not always consistent ways, with the result that the tradition on this point is confused.[1] There is, however, general agreement on the simpler usage of the three technical names **difference, property** and **accident,** and the student will find it helpful to be acquainted with them.

Any quality or attribute is regarded as being either a difference or a property or an accident.

A quality is said to be a **difference** if it serves to distinguish the class of entities of which it is a quality from other species of the same genus, i.e. if it is utilized in the definition of the class.

[1] Cf. the similar situation, and the reasons for it, mentioned on page 22.

A quality is said to be a **property** if it is a quality necessarily possessed by every member of the class, yet not utilized to distinguish the class from other species of the same genus.

A quality is said to be an **accident** if it may indifferently belong or not belong to all or any of the members of the class.

To illustrate, consider the members of the present Cabinet, all of whom have the qualities or attributes of being Cabinet Ministers and of being Privy Councillors, while some of them have the quality or attribute of being under forty years of age.

Their being *Cabinet* Ministers is a difference, as it serves to distinguish them from the other species of the same genus, namely Ministers not in the Cabinet.

Their being *Privy Councillors* is a property, for all Cabinet Ministers are Privy Councillors (the Cabinet being technically a committee of the Privy Council).

Their being under forty years of age is an accident, for their age is irrelevant as far as their being Cabinet Ministers is concerned.

Before leaving this topic, the last of the traditional formal logic to be discussed, it ought to be noticed that the decision whether a specified quality in a given case is a difference or a property or an accident depends to some extent on one's point of view; just as any scheme of division depends to some extent on one's point of view.

That last observation recalls the wider comment with which we began and with which it is well to end, namely that there are many points of view on any logical question, and that the traditional formal logic represents only one of them. However, it is for many reasons an important one, and at the present day, after more than two thousand years of varying influence, the traditional doctrine still survives as part of a liberal education and as prolegomena to any further inquiry.

During its long descent the tradition has from time to time accumulated corollaries, developments and refinements, in quantities shown by the excessive bulk of the common text-books, but all these elaborations fall within the system outlined here, and the reader who is reasonably familiar with the matter of the foregoing chapters knows enough of the traditional formal logic to profit by the discussion of its topics in any wider or different context.

THE TRADITIONAL PROPOSITIONAL FORMS

THE A PROPOSITION

The Universal Affirmative ; *All S are P* ; *S a P*.
S is distributed
P is undistributed, $\bar{S} a \breve{P}$.

Inclusion of S in P,

or Coextension of S and P,

THE E PROPOSITION

The Universal Negative ; *No S are P ; S e P*.
S is distributed,
P is distributed, $\bar{S} e \bar{P}$

Exclusion of S and P,

THE I PROPOSITION

The Particular Affirmative ; *Some S are P* ; *S i P*.
S is undistributed,
P is undistributed, $\breve{S} i \breve{P}$.

Intersection of S and P,

or Inclusion of S in P,

or Inclusion of P in S,

or Coextension of S and P,

THE O PROPOSITION

The Particular Negative ; *Some S are not P ; S o P*.
S is undistributed,
P is distributed, $\breve{S} o \bar{P}$.

Intersection of S and P,

or Exclusion of S and P,

or Inclusion of P in S,

If a proposition in *S-P* form is true, certain other propositions in *S-P* form are true also. I.e. the latter can be inferred from the former.

SUBALTERN RELATION

$\check{S} i \check{P}$ is subaltern to $\check{S} a \check{P}$

$\check{S} o \check{P}$ is subaltern to $\check{S} e \check{P}$.

If a proposition in *S-P* form is true, certain other propositions in *P-S* form are true also. I.e. the latter can be inferred from the former.

CONVERSE RELATION

Converse of $\check{S} a \check{P}$ is $\check{P} i \check{S}$

Converse of $\check{S} e \check{P}$ is $\check{P} e \check{S}$

Converse of $\check{S} i \check{P}$ is $\check{P} i \check{S}$

Converse of $\check{S} o \check{P}$ *NONE*.

If a proposition in *S-P* form is true, certain other propositions in the form *S not-P* are true also. I.e. the latter can be inferred from the former.

OBVERSE RELATION

Obverse of $S a P$ is $S e$ *not-P*.

Obverse of $S e P$ is $S a$ *not-P*.

Obverse of $S i P$ is $S o$ *not-P*.

Obverse of $S o P$ is $S i$ *not-P*.

Note.—Terms that are undistributed in the original proposition must remain undistributed in the derivative proposition if the immediate inference is to be valid.

If a proposition in *S-P* form is true, certain other propositions in *S·P* form are false.

CONTRARY RELATION

    S a P and *S e P* are Contraries.

CONTRADICTORY RELATION

    S a P and *S o P* are Contradictories.
    S e P and *S i P* are Contradictories.

THE SQUARE OF OPPOSITION

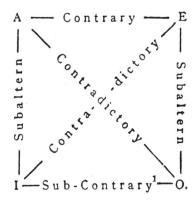

[1] This name for the relation between the particular affirmative and the particular negative was invented mainly to complete the mnemonic diagram. That is why it is mentioned only here, and not in the chapter on immediate inference.

TABLE OF IMMEDIATE INFERENCE

Given Proposition	Subaltern	Converse	Obverse	Contrary	Contradictory
S a P	*S i P*	*P i S*	*S e not-P*	*S e P*	*S o P*
S e P	*S o P*	*P e S*	*S a not-P*	*S a P*	*S i P*
S i P	—	*P i S*	*S o not-P*	—	*S e P*
S o P	—	—	*S i not-P*	—	*S a P*

SYLLOGISM

THE THREE TERMS INVOLVED

S The Minor Term, appearing in the Conclusion as Subject.

P The Major Term, appearing in the Conclusion as Predicate.

M The Middle Term.

FIGURES OF THE SYLLOGISM

1. *M P*	2. *P M*	3. *M P*	4. *P M*
S M	*S M*	*M S*	*M S*
∴ *S P*	∴ *S P*	∴ *S P*	∴ *S P*

The four Figures can be memorized by the following diagram :—

1. ⊇   2. ⊐   3. ⊏   4. Z

THE THREE RULES OF DISTRIBUTION IN SYLLOGISM

1. *S* and *P* must be undistributed in the Conclusion if undistributed in the Premisses.
2. *M* must be distributed once at least.
3. One premiss at least must be affirmative.

FORMAL FALLACIES

Illicit Process of the Minor⎫ (Contravention of Rule 1)
Illicit Process of the Major⎭

Undistributed Middle   (Contravention of Rule 2)

HYPOTHETICAL ARGUMENT

(Note that p and q are symbols for *propositions*, not for *terms*.)

If p, then q
p
———
∴ q Affirming the Antecedent. Valid.

If p, then q
not-q
———
∴ not-p Denying the Consequent. Valid.

(To deny the antecedent, or to affirm the consequent, is invalid.)

———————

DISJUNCTIVE ARGUMENT

(a) Weak disjunction

p or q
not-p
———
∴ q Denying an Alternative. Valid.
(To affirm an alternative is invalid.)

(b) Strong disjunction

(i) p or q (*but not both*)
not-p
———
∴ q Denying an Alternative. Valid.

(ii) p or q (*but not both*)
p
———
∴ not-q Affirming an Alternative. Valid.

INDEX

For Product Safety Concerns and Information please contact our EU
representative GPSR@taylorandfrancis.com
Taylor & Francis Verlag GmbH, Kaufingerstraße 24, 80331 München, Germany